ALUMINIUM IN BUILDING

Aluminium in Building

John Lane

ASHGATE

Published by
Ashgate
Ashgate Publishing Limited
Gower House
Croft Road
Aldershot
Hants GU11 3HR
England

Ashgate Publishing Company
Old Post Road
Brookfield
Vermont 05036
USA

A CIP catalogue record for this book is available from the British Library and the US Library of Congress.

ISBN 1 85742 082 9

Typeset in 11 point Baskerville by Poole Typesetting (Wessex) Ltd., Dorset and printed in Great Britain at the University Press, Cambridge.

Contents

List of tables

List of figures

Acknowledgements

The author acknowledges the help provided by the manufacturers of many products referred to in this publication, and by the many companies who have assisted by providing photographs.

To illustrate up-to-date practices the author has drawn upon the publications and sales literature of many companies, and illustrations from those publications have been reproduced here.

In particular, thanks are due to:

Alumasc
Aluminium Federation
Aluminium Window Association
Avdon
Baco Contracts
British Alcan Building Products
British Alcan Extrusions
British Alcan Rolled Products
Caradon Everest
Crittall Windows
Finalex
Glostal
Heywood Williams
Hydro Aluminium Century
Kawneer Europe
Kaye Aluminium
Pearce & Cutler

Smart Systems
Stoakes Systems
Tru Architectural Products

The author also acknowledges all the information and help received over many years by members of companies in the aluminium industry, which has provided the strong technical background without which this book would not have been possible.

Introduction

Aluminium is a relatively young metal. It was in 1807 that Sir Humphry Davy first suspected the existence of the metal, but was unable to isolate it. It was first isolated a few years later, in 1825, by H. C. Oersted in Denmark, and by 1854 a Frenchman, Henri Claire De Ville, had developed a chemical extraction method using sodium that enabled small quantities of the pure metal to be obtained from its oxide. At this time aluminium was both a novelty and a 'precious' metal with its value rivalling that of silver and gold. Indeed, the metal was so esteemed that Napoleon III of France had a dinner service made from it.

Then in 1886 came the commercial breakthrough. Quite coincidentally, two scientists working independently, Charles Hall in the USA and Paul Héroult in France, discovered the electrolysis process for extracting aluminium from its oxide, alumina. The process required significant quantities of electricity and with major sources of electricity being established in the same era commercialization of the electrolysis process, called Hall–Héroult in honour of the two inventors, became a practical proposition. Rapid development followed with the metal's price falling from £1 000/tonne down to less than £100/tonne by the turn of the century.

The building industry was one of the first to appreciate the merits of the lightweight, durable and silvery-looking metal. In 1897, the cupola of San Gioacchino Church in Rome was covered with aluminium sheeting, and five years later a Roman synagogue was also clad with the metal. Both are still giving good service. An even earlier, and more well-known example is the statue of Eros in Piccadilly Circus, London. This cast construction, erected in 1893, has since been renovated and found to be in excellent condition.

Today the building industry is one of aluminium's major markets, using alloys and finishes undreamt of in the early days, and utilizing casting, rolling and extrusion processes to obtain a multitude of forms and shapes for applications ranging from curtain rails and carpet edging to curtain walling and space structures.

With about twenty million tonnes now being produced annually, aluminium has become the second most widely used metal after steel, and is used for applications that have helped to revolutionize modern society. For example, high-speed travel would not be possible without the significant usage of aluminium in the construction of all kinds of transport, and modern, high-speed printing presses would be ineffective without the use of specially treated aluminium for litho plates.

The first part of this publication is a general discussion of the metal covering aspects such as properties, alloys, fabrication, joining, finishing and durability.

Then the characteristics of aluminium that are of particular importance in building are discussed and the final part highlights some of the major uses of aluminium in the building industry, as a demonstration of the metal's versatility and range of applications.

Part I
THE METAL

1 Obtaining the metal

The solid matter of the earth's crust is made up of nearly a hundred elements, the most abundant of which are oxygen, silicon, and then aluminium. So aluminium is the third most common element and in the form of various compounds the metal is very widely distributed. Minute traces of aluminium salts are found in many foods for example, and the element is one of those around which evolution on this planet has developed. Aluminium, unlike some other metals, does not occur naturally in its pure metallic form. Only certain of the multitude of aluminium-bearing minerals may be regarded as ores, since extraction from many of them is currently too uneconomical. Commercial ores are grouped under the generic term 'bauxite', derived from the medieval village of Les Baux in Southern France, where high concentrations of hydrated aluminium oxide are found, and where early mining of the ore took place.

There are widespread sources of bauxite throughout the world, including in addition to Les Baux, Greece, Hungary, Jamaica, Africa, the USA, Australia and South America. It is a metal in plentiful supply.

The reduction of bauxite to metal involves two main operations. First the bauxite is treated chemically to remove impurities and obtain aluminium oxide, alumina (Figure 1.1(a)). Then using the Hall–Héroult electrolysis process the alumina is reduced to aluminium, that is to say, the oxygen is removed. The basic process discovered by Hall and Héroult is still the one used today, although refinements in the techniques used have resulted in much more power-efficient operations. The process involves a bath of fused cryolite containing dissolved alumina which is electrolysed

Mining of bauxite takes place in many parts of the world

by the passage of a high amperage, low voltage current between carbon anodes and the carbon lining of the cell, which forms the cathode. The alumina is decomposed into oxygen and aluminium. The former is liberated at the anodes and the pure aluminium sinks to the bottom of the cell from where it is tapped off periodically and cast ready for further treatment (Figure 1.1(b)). The production process operates with many of these electrolytic cells being connected in series of lines (known as pot lines) of up to 150 in number. Continuous operation is maintained as a necessity to avoid the cooling down and subsequent solidifying of the baths.

Cheap electrical power is the key to the production of low-price, high-grade aluminium. Hydroelectric power was the obvious first source, and the aluminium industry has its origins in areas like Switzerland, Scotland, Canada and Norway where water power was available in abundance to be harnessed for the production of electricity.

Today, hydropower remains a major source of electricity for aluminium extraction, and one which is completely acceptable environmentally, while in addition coal, natural gas and nuclear energy are also utilized. In the UK for example, there are smelters operating using electricity generated from hydropower, coal, and nuclear energy.

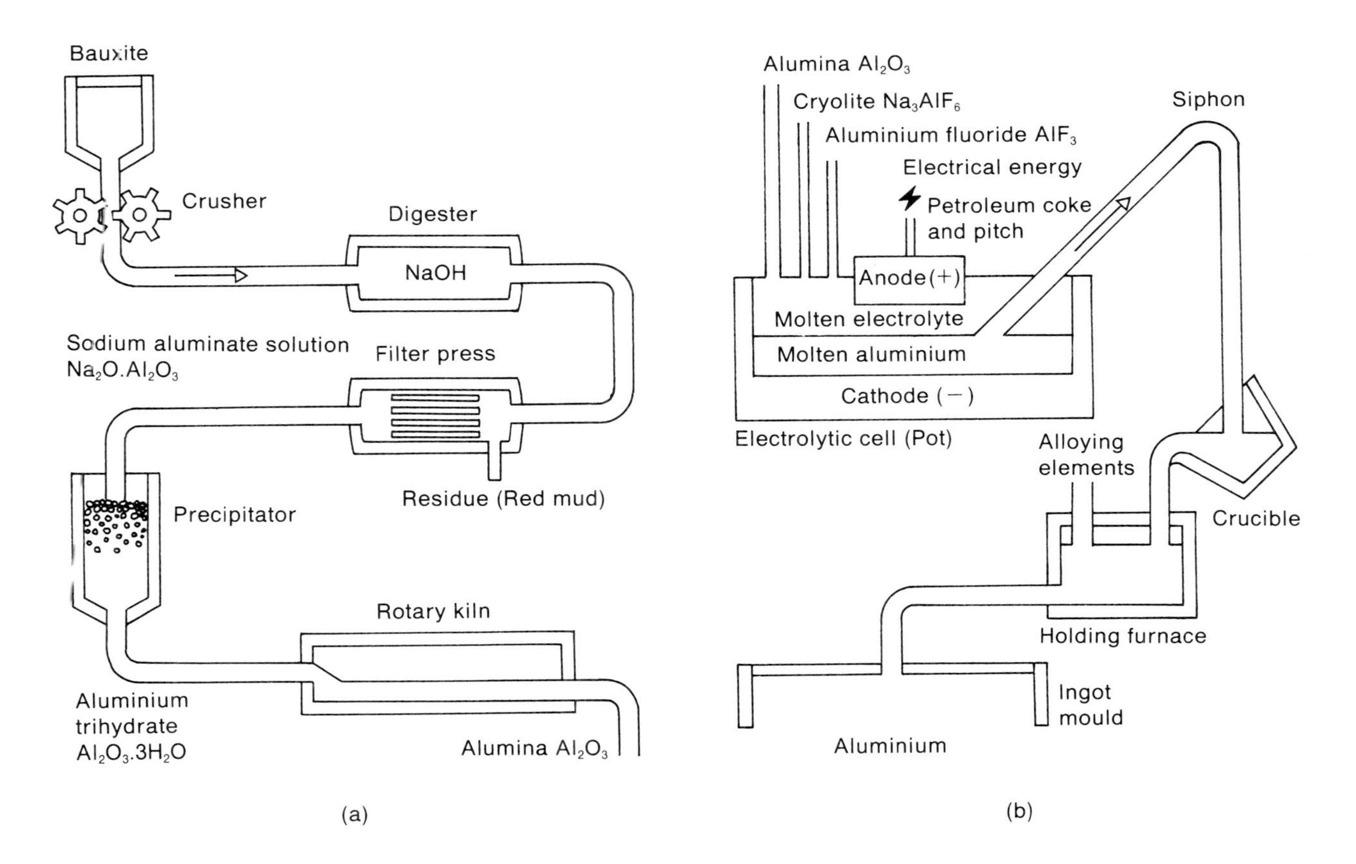

Figure 1.1 **From bauxite to aluminium: (a) stage 1 – the chemical process; (b) stage 2 – the electrolytic process**

General view inside an aluminium smelter showing the lines of 'pots' in which the alumina (aluminium oxide) is turned into pure aluminium

Hydroelectric power has been used in the Scottish highlands by British Alcan since 1894. The Laggan dam, seen here, is part of a project that feeds water to British Alcan's Lochaber power station at the foot of Ben Nevis. First opened in 1929 the power station supplies the adjacent Lochaber smelter, which produces around 40 000 tonnes of aluminium annually

The Lochaber smelter of British Alcan at the foot of Ben Nevis. The pipes carrying the water down the mountainside to the power station are clearly visible

2 Properties of the metal

Aluminium's unique properties have set the metal apart as a special building material exhibiting outstanding properties: strength, lightness, durability and versatility. There are many other properties of aluminium that make it particularly suitable for specific applications outside building – its electrical properties make the metal ideal for busbar and overhead cables; its heat conductivity makes it ideal for car radiators and hollow-ware; its malleability and compatibility with foodstuffs provide many uses in packaging, and its surface treatment characteristics make aluminium the major material for lithographic printing plates. But in the construction industry it is the combination of strength and light weight, the high natural resistance to atmospheric attack, and the ease with which the metal can be formed that have led to such a wide range of vastly different uses for aluminium in building.

Additionally the ease with which the metal's surface can be treated in various ways to provide a range of decorative and protective finishes adds to aluminium's list of advantages.

Alloys

Up to this point reference has been made only to 'pure aluminium' and 'the metal'. In fact the term 'aluminium' is widely used generically to encompass a whole family of alloys – each with its own specific properties and applications.

Table 2.1 Properties of some typical casting alloys used in building

Alloy BS 1490	Condition	Mechanical properties: Minimum tensile strength (MPa) Sand	Chill	Elongation (min) % Sand	Chill	Casting process and applications
LM2 (Al–Si 10 Cu2–Fe)	M	–	150	–	–	Pressure diecastings General purpose
LM4 (Al–Si 5–Cu3)	M T6	140 230	160 280	2 –	2 –	Sand, permanent mould and pressure diecasting General purpose
LM5 (Al–Mg 5)	M	140	170	3	5	Sand and permanent mould. Takes good anodic finish
LM6 (Al–Si 12)	M	160	190	5	7	All processes Good castability and corrosion resistance
LM24 (Al–Si 8–Cu3–Fe)	M	–	180	–	1.5	Pressure diecastings General purpose similar to LM2
LM25 (Al–Si 7–Mg)	M T6	130 230	160 280	2 –	3 2	Sand and permanent mould castings Offers good mechanical properties
LM27 (Al–Si 7–Cu2)	M	140	160	1	2	All processes Similar to LM4

Pure aluminium easily alloys with many other elements. Among these, magnesium, silicon, manganese, copper and iron are regularly used. More recently lithium has been added to the range providing alloys of much higher strength and lower density than any of the traditional alloys. These Al-lithium alloys are of particular value for the aircraft industry but are not expected to be used for general engineering applications owing to their high cost of production.

Alloy classification

Aluminium and its alloys are divided into two broad classes, cast and wrought. The latter class is subdivided into non heat-treatable and heat-treatable alloys. In the non heat-treatable group, properties are altered by

Table 2.2 Properties of some typical wrought alloys used in building

Alloy	Condition*	Mechanical properties**				Form and applications
		0.2% proof stress	Tensile strength minimum MPa	Tensile strength maximum MPa	Elongation on 50 mm minimum %	
1080 A (99.8% Al)	O	–	–	90	29	Sheet. Very malleable Flashings
1200 (99.0% Al)	O H8	– –	70 140	105 –	30 4	Sheet and extrusions Standard commercial purity General purpose
3103 (Al–Mn1)	O H8	– –	90 175	130 –	24 4	Sheet. Good all purpose roofing sheet, particularly profiled
5251 (Al–Mg2)	O H6	60 175	160 225	200 275	18 5	Sheet and extrusions Improved strength and good durability
5083 (Al–Mg4.5–Mn)	O H4	125 270	275 345	350 405	16 8	Sheet and extrusions Structural alloy. Good in marine environments
6063 (Al–Mg–Si)	T4 T6	70 160	130 185	– –	14 7	Extrusions Most common general purpose extrusion alloy
6082 (Al–Mg–Si–Mn)	T4 T6	120 270	190 310	– –	14 8	Sheet and extrusions Good structural alloy

* See Tables 3.1 and 3.2 for description. Other tempers and conditions may also be available for some alloys.

** Properties depend on thickness and may vary from figure shown.

the degree of cold-working that is performed, such as rolling. In the heat-treatable group, strength is affected by the application of various heat-treatments.

Some casting alloys are also strengthened by heat-treatment.

Alloy designations

Castings

Casting alloys for general engineering applications are specified in BS1490, aluminium alloy ingots and castings. They are numbered from 0

Table 2.3 Typical characteristics and processing suitability of typical wrought alloys

Material designation and temper		Resistance to atmospheric attack	Formability	Suitability for TIG/MIG welding	Suitability for anodizing
1080 A	O	E	E	E	E
1200	O	V	E	V	V
3103	O	V	E	V	G
5251	O	V	V	V	V
5083	O	V	G	E	V
6063	T4	V	V	V	V
6082	T4	V	V	V	V

E = Excellent V = Very good G = Good

to 30 and prefixed with the letters LM (originally meaning light metal). Not all of the numbers in the sequence are now in use. Various suffixes are also employed to indicate the condition or heat-treatment condition of the alloy.

Wrought

Wrought alloys are specified by a series of British Standard specifications. They are classified by chemical composition in an internationally agreed four-digit coding system. The first of the four digits indicates the alloy group as decided by the major alloying elements included, and are as follows:

1xxx Aluminium of 99.00% minimum purity or higher
2xxx Copper
3xxx Manganese
4xxx Silicon
5xxx Magnesium
6xxx Magnesium and silicon
7xxx Zinc
8xxx Lithium and others
9xxx Unused series

The remaining three digits are available to indicate alloy modifications.

Wrought alloys are further subdivided into non heat-treatable (that is work-hardening) and heat-treatable alloys.

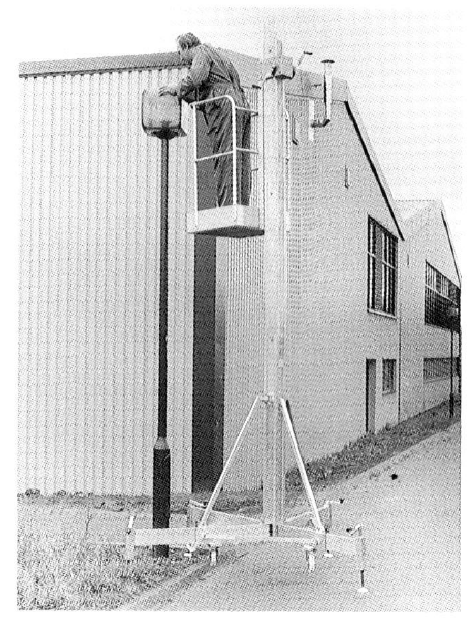

This light-weight access platform utilizes two specially designed telescopic sections. The system is easily transported, assembled and dismantled (courtesy Skywinder)

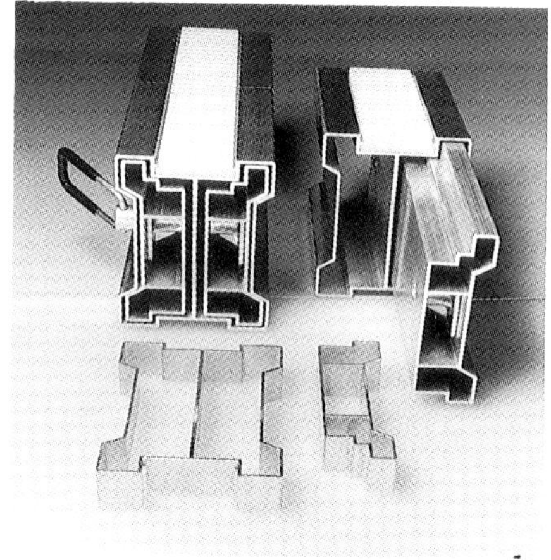

Close-up of the two main telescopic hollow sections used in the Skywinder access platform

Specific properties

Weight

Pure aluminium is one of the lighter elements with a density of 2.7 g/cm^3 (0.098 lb/in^3). The densities of its alloys vary, most of them fall within the relatively narrow band of 2.63–2.80 g/cm^3, while the lithium-based alloys have a density of around 2.55 g/cm^3.

Those alloys with densities lower than that of pure aluminium, apart from the lithium ones, are those in the Al–Mg series, due to magnesium being lighter than aluminium.

Strength and ductility

As a very rough guide, strength and ductility are inversely proportional. When one is high the other is low. Where severe forming is to be carried out, for example on a roof flashing, then the material being used (for

Testing a metal sample for its tensile properties

example commercial-purity sheet) should be specified in the soft, ductile condition designated by the alloy suffix 'O'.

Tensile strength is not significantly affected by temperature fluctuations in atmospheric conditions, but if very high temperature service conditions are anticipated then it should be noted that aluminium alloys begin to lose strength slightly above 100°C. Above 200°C there is a more significant drop, so advice should be sought if these relatively high operating temperatures are likely to be encountered.

Unlike mild steel, aluminium does not have a sharply defined yield point, and so a proof stress figure is normally used for design calculations. The 0.2 per cent proof stress is defined as that stress which produces, while a load is still applied, a non-proportional extension equal to 0.2 per cent of the gauge length.

Elasticity

Young's modulus of elasticity for aluminium is about one-third that of steel. This means that an aluminium structural member under load, and having the same section properties as a steel member will deflect to a greater extent than the steel one. The impact absorption properties of the aluminium are greater than those of an equivalent steel member.

To obtain equivalent or better rigidity than with steel it is common

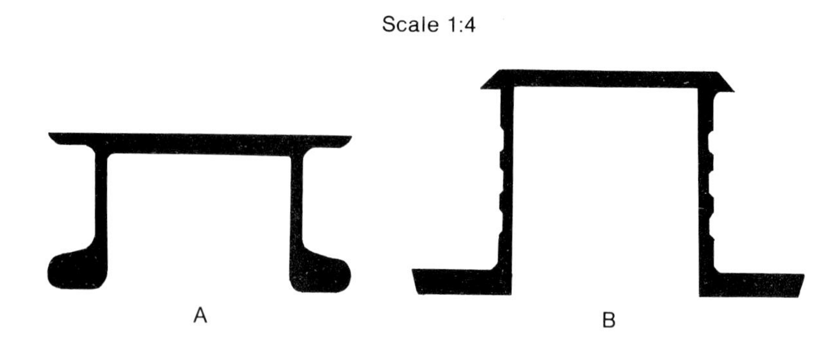

Figure 2.1 Aluminium 'copy' of a steel design for a trench shoring section used in the construction industry (A); redesign to maximize the structural characteristics (B)

practice with aluminium sections to make them thinner and deeper than their steel counterparts. By designing to take advantage of the different characteristics of aluminium compared with steel the best results can be obtained (Figure 2.1).

As a general approximation it can be said that if an aluminium section is designed to have the same stiffness as the steel section it replaces, its weight will be one-half that of the steel section, rather than one-third if an exact copy were made.

Corrosion resistance

The aluminium oxide 'skin' that is always present on the surface of aluminium exposed to oxygen provides both excellent atmospheric durability and resistance to a large number of chemicals. It is this excellent durability that is one of the main reasons why aluminium is so widely used throughout the building and construction industry. This ever-present oxide skin may be thickened electrolytically to give enhanced protective properties; the electrolytic process, known as anodizing, is extensively used commercially to improve the durability of aluminium building components and products. It is also used to provide distinctive decorative colour finishes.

Thermal expansion

Pure aluminium has a coefficient of linear expansion of $24 \times 10^{-6}/°C$ over the range 20–100°C. Alloying elements have only a very small effect upon the value of this coefficient. This means that the dimensional increase with a given rise in temperature is about twice that of steel but only one-third that of polyvinyl chloride (pvc), for example. Such differences in expansion and contraction need to be considered where aluminium is used in bimetallic or composite constructions, and the relative expansions of alternative materials may well be a deciding factor in the selection of material for a given application – for example the choice of aluminium versus pvc for window frames or guttering in hot climates.

Conductivity

Pure aluminium has a heat conductivity value of 230 W/m/°C which gives the metal a conductivity 60 per cent that of copper and 450 per cent that of mild steel. This good heat conductivity is exploited in many applications, ranging from the long-standing traditional example of hollowware to industrial heat-exchangers and automotive radiators.

Thermal conductivity reduces with increased alloying of aluminium.

Reflectivity

A bright aluminium surface reflects about 75 per cent of the light and 90 per cent of the heat radiation that falls upon it. These radiation properties are not significantly lowered by the natural surface dulling and oxidizing that can occur over the passage of time.

These reflectivity properties, along with the metal's impermeability are being put to good use as thermal insulation barriers in domestic, office and industrial buildings.

3 Fabrication of aluminium alloys

Virgin aluminium produced in the smelting process may be converted into a number of semi-fabricated forms. These are rolled products, including plate, sheet and foil; other rolled products such as bar, rod and wire; extruded products, including solid and hollow sections and tubes; castings, either sand, gravity die or pressure die; forgings; impact extrusions; and paste.

The major product groups of interest to the building industry are briefly described in the following sections.

Rolled products

The definitions of rolled products are as follows:

Plate: Flat material, either hot or cold rolled, over 6 mm in thickness.
Sheet: Cold rolled material, either flat in thicknesses over 0.2 mm but not exceeding 6 mm, or coiled in thicknesses over 0.2 mm but up to 3 mm only.
Foil: Cold rolled material in thicknesses of less than 0.2 mm.

Rolling aluminium and its alloys is one of the major ways of converting cast primary aluminium from the smelters into a usable industrial form.

By rolling, it is possible to reduce an ingot of primary metal down to plate material having a maximum thickness of around 250 mm, right down to the thinnest of foil thicknesses, sometimes as little as 0.006 mm, approximately one-third the thickness of a human hair.

The rolling of cast aluminium changes its metallic structure and the metal takes on new characteristics and properties. The brittleness of the coarse cast structure is replaced by a stronger and more ductile material, with the degrees of strength and ductility being variable factors that are functions of the amount of rolling given to the metal, the rolling temperature and the alloy composition.

Rolling of aluminium consists of two basic stages. First a slab, or ingot, of aluminium weighing up to 12 tonnes is hot rolled by repeated passes through a 'breaking-down' mill followed by a pass through a series of mills operating in tandem in order to obtain a coil of metal having a thickness of approximately 5 mm. The second stage consists of a series of cold rolling operations taking the metal to its final thickness.

Hot rolling takes place using ingot that has been scalped on each side to remove oxide and surface roughness and then heated in special furnaces to a temperature of around 500°C. At this temperature, which varies slightly according to the alloy being rolled, the metal is malleable enough to withstand severe thickness reduction per pass through the mill without undergoing any work-hardening. By hot rolling, an ingot having a thick-

Handling an aluminium ingot prior to hot rolling

Hot rolling an aluminium ingot at the Rogerstone Works of British Alcan Rolled Products

Twelve-tonne aluminium coils on the hot-mill tandem at British Alcan Rolled Products

ness of, say, 380 mm can be reduced down to a coiled length having a thickness of 5 mm without any recourse to intermediate annealing (softening). Plate material, defined as over 6 mm thick, and often many times this thickness is almost entirely produced by hot rolling only. It is only the thinnest of plate material, approaching the upper thickness limits of sheet, that is sometimes given a final cold rolling pass.

All sheet and coil material undergoes a series of passes through cold mills in which the metal thickness is progressively reduced. During these passes cold-working of the metal takes place resulting in metal hardening. This phenomenon is exploited in those alloys known as 'work-hardening' alloys to obtain sheet and coil having differing hardness and ductility. Thus not only do mechanical properties vary from alloy to alloy but also within a single alloy specification by altering the work-hardening content.

A further group of alloys, those known as heat-treatable, gain their strength from a heat-treatment process after final rolling. Table 3.1 shows the effect of increased hardness on strength of the alloy.

Table 3.1 Temper description

Description	UK designation	Minimum tensile strength MPa 1200 alloy*
Annealed, fully soft	O	70
As manufactured	M	–
Quarter-hard	H2	95
Half-hard	H4	110
Three-quarters hard	H6	125
Fully-hard	H8	140

*Shown as an illustration of the effect of increasing temper on strength.

Aluminium foil is produced in a similar manner to sheet using specialized mills that enable the metal thickness to be reduced to the finest of webs. A particular characteristic of foil rolling is that of 'double rolling' where two layers of metal are rolled together with touching surfaces. This technique enables fast, accurate rolling of very thin foil to take place with a minimum risk of web breakage under the tensions applied during rolling. The process results in the characteristic appearance of foil having one surface bright and the other matt.

During the rolling operation it is possible to introduce patterns and surface texture into the metal, by using special embossed rolls. By this means heavily embossed plate, known as tread-plate, which is popular in many building and construction applications, patterned sheet for a variety of decorative uses from tea trays to kick plates, and textured foil for

High-speed cold rolling of aluminium foil

many decorative and functional uses in applications from packaging to insulation can be produced.

Extruded products

The versatility of aluminium as a metal is complemented by the design scope offered by the extrusion process. Other metals can be extruded but none with the ease of aluminium and its alloys. Plastics materials can be extruded into shapes just as complicated, but the tooling involved is considerably more complex and expensive. Relatively speaking the costs for producing aluminium extrusion dies are inexpensive and so the production of special shapes for specific individual applications is commonplace. This is evidenced by the many thousands of extrusion dies held by UK extruders.

The ductility of aluminium in a hot state enables an unlimited variety of shapes to be produced. Not all shapes imaginable can be extruded as there are certain production parameters that must be observed, but within these limitations the range of shape complexity that can be achieved is considerable. The availability of the extrusion process has, over the past few decades, revolutionized the design approach to many products and problems. Today's aluminium window frames and commer-

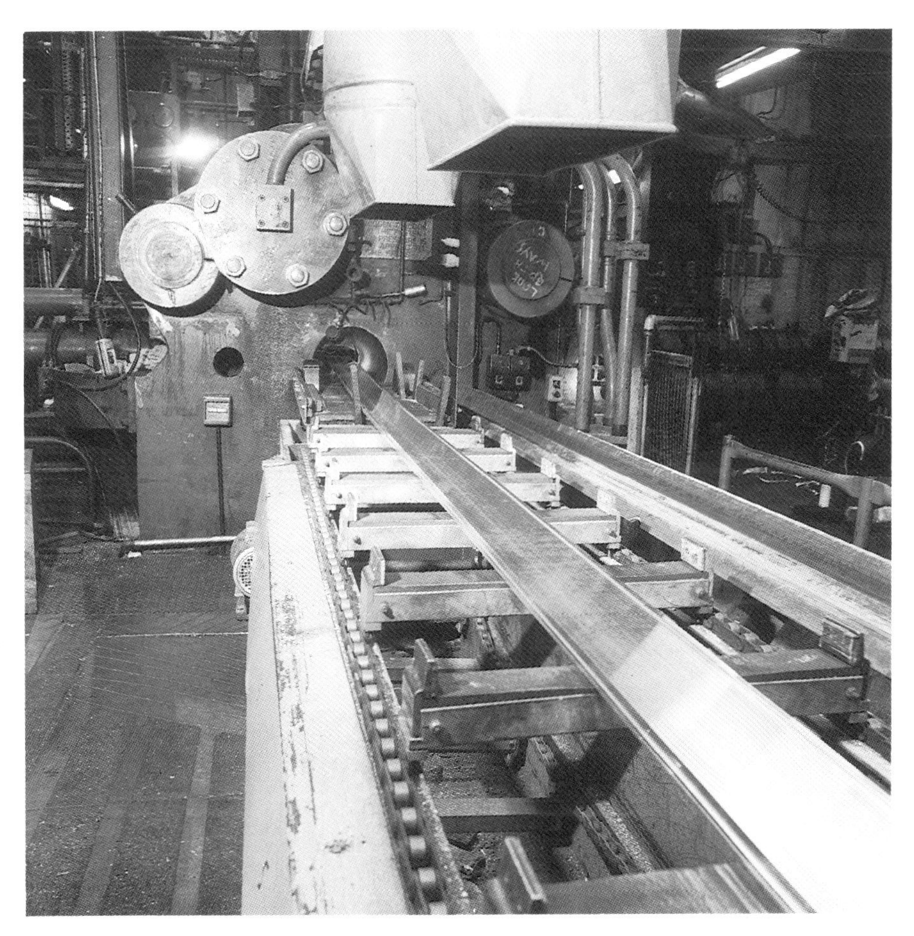

An aluminium extrusion leaving the mouth of the press during the extrusion operation (courtesy Kaye Aluminium)

cial curtain walling sections are typical examples. Designing with aluminium extrusions is different from designing with other metals or with wood. Because so many features can be built into a single extruded shape, fabrication and assembly of components is simplified and design scope is greatly increased.

Put in its most simplistic form, extrusion may be compared to the domestic operation of squeezing icing through a syringe, where the shape of the squeezed-out icing is determined by the shape of the nozzle orifice. Aluminium extrusions are made by forcing hot metal through a specially shaped opening in a steel die. Blocks of cylindrically-shaped aluminium are heated to a temperature of around 500°C and then inserted in a

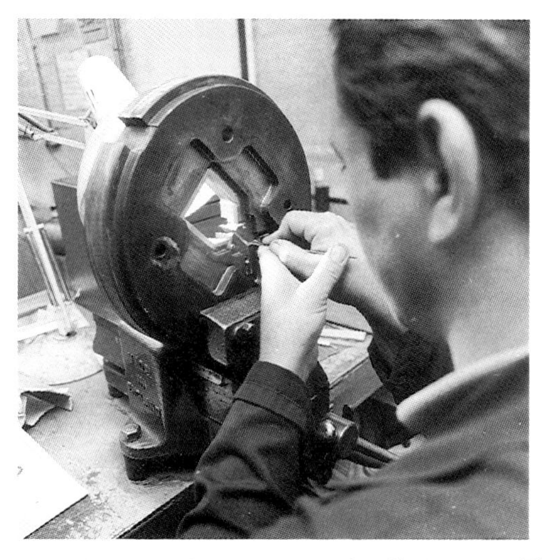

Putting the finishing touches to an extrusion die (courtesy Finalex)

These two extrusions demonstrate the way in which multiple features can be incorporated into an extrusion design. Here we have a hinge fit, clip fit, decorative grooving and channels

container in a hydraulic press. The ductility of the hot metal enables it to be forced through very complex shapes both solid and hollow, producing sections of incredible complexity with relative ease.

Most extrusion alloys fall in the heat-treatable category, which means that the strength of an extruded section is determined by its alloy specification and by the heat-treatment processing it receives after extrusion.

Significantly, for some alloys the temperature and time required for solution heat-treatment corresponds with the temperature of extrusion

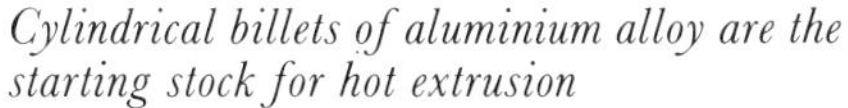

Cylindrical billets of aluminium alloy are the starting stock for hot extrusion

General view of an extrusion bay showing aluminium extrusions on the run-out table (courtesy Finalex)

and duration of extrusion time. This means that in effect, these particular alloys may be considered to be in the first stage of heat-treatment as the metal leaves the extrusion die. The second stage of heat-treatment, that of rapid quenching can be, and is, accomplished by forced-air cooling or water immersion at the press exit. Thus the most commonly used medium-strength architectural alloys, such as 6063 (Al–Mg–Si), are heat-treated as part of the extrusion operation, providing an extremely effective process both from metallurgical and cost considerations.

Some other alloys, such as 6082, need to be 'aged' to provide a full heat-treatment. This final operation is carried out at relatively low temperatures (around 170°C) in special ageing ovens, or in some cases is allowed to take place at ambient temperature over an extended period of time.

Annealing, of course, is a type of heat treatment. It is a process used to soften the aluminium to obtain maximum ductility. By contrast the other heat-treatment processes are used to strengthen the metal and the terminology of 'heat-treatment' is accepted as applying to the process of strengthening rather than softening (Table 3.2).

Castings

Casting is one of the oldest processes for producing articles in metal. There are many excellent aluminium alloys that have been specially designed for casting using processes from sand casting to gravity die- and pressure die-

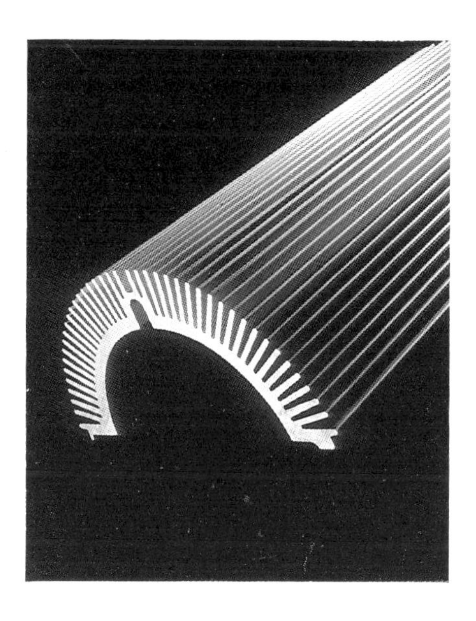

Standard shapes as well as complex specials are available

Table 3.2 Heat-treatment designations

The following symbols are used in the UK as suffixes to the alloy designation to denote the condition of the metal specified:

Description	Designation
Annealed, soft	O
As manufactured (direct from the press with no special treatment)	F
Solution heat treated and naturally aged	T4
Cooled and artificially aged	T5
Solution heat treated and artificially aged	T6

casting. The alloys are quite different in composition from wrought alloys, and have been formulated to give the specific qualities of metal flow and cooling required to make high-quality castings.

There are three main groups of casting alloy:

Aluminium–silicon
Aluminium–copper
Aluminium–magnesium

The aluminium–silicon alloys are very popular. They combine good castability, pressure tightness and resistance to corrosion with a good range of mechanical and physical properties. The range of silicon content is from around 2 per cent to 13 per cent depending upon the specification. Automotive and industrial components, and various household appliance parts are typical end uses. Anodizing can be carried out but the finish obtained darkens as the silicon content increases and it is not possible to obtain a good colour match with anodized wrought components.

The aluminium–copper alloys are more susceptible to a defect known as hot-shortness, which is overcome by correct foundry technique, but have the advantage of offering excellent machining characteristics. Many stressed parts are made of these alloys.

The third group is that of the aluminium–magnesium alloys. These require more careful foundry techniques particularly to minimize the tendency to oxidation. However, the group provides alloys of high strength, excellent corrosion resistance and finishing characteristics. Such alloys therefore are well suited for architectural components where durability and good appearance are required.

There are three processes of casting in common practice. These are sand, permanent mould and die-casting.

Moulds made from sand are used where the number of pieces to be made does not justify the expense of making metal moulds or where the size of casting involved is very large. Sand moulds are fed with molten metal through carefully placed 'runners' with the avoidance of air pockets being a prime consideration. 'Risers' are included in the mould to allow molten metal to 'escape' and to provide reservoirs of hot metal to offset shrinkage.

Permanent cast-iron moulds give a more economical process both because of the higher production rate that is possible and because of the closer dimensional accuracy and smoother surfaces that are obtained, which reduce after-casting finishing. Metal moulds are generally fed by gravity but forced flow of the molten aluminium is sometimes achieved by low-pressure injection.

Die-casting involves the injection of molten metal under pressure or vacuum into a steel die. It is a high-speed process well suited to the economic production of small castings in high volume.

The alloys used for die-casting are not suitable for subsequent heat-treatment, and because of the tendency to porosity the method is not used for castings requiring to be highly stressed. The close dimensional accuracy that is achieved means that castings require very little subsequent machining before use.

Wire and bar

Bar is defined as round, rectangular or polygonal solid section supplied in straight lengths. Such material is not less than 6 mm diameter or width across flats.

Wire is material of similar shape, but no more than 10 mm diameter or width across flats. It is usually supplied in coil form.

Wire and bar are produced by first hot and then cold working. In the hot rolling stage heated rod is passed between a series of grooved rolls, taking the metal diameter down to around 10 mm, after which further reductions are made by cold drawing. Hot rolling of bar is a high-speed, low-cost method of producing large quantities of bar in a selected alloy and size. Bar may also be produced by extrusion in a similar manner to the production of other extruded shapes.

Wire is produced from hot rolled bar by high-speed cold drawing, followed by coiling.

Various alloys are used for wire and bar. These include electrical conductor alloy, high-strength, heat-treatable alloys for rivets and small stressed parts (particularly in aircraft), machining stock, various welding and brazing filler alloys, and general-purpose mechanical engineering materials.

Forgings

Developed to highly sophisticated modern production from the elementary craft of the blacksmith, forging is a process that can produce aluminium alloy components of high metallurgical properties. The basic production practice is that of die-forging where the aluminium is hammered or squeezed between a set of steel dies machined to the exact shape required and highly polished.

Forging reached its zenith when propeller-driven aircraft were at their peak and forged propeller blades were made in large quantities.

Coining is a supplementary process to die-forging, carried out when high dimensional accuracy is required on small forgings. The production of aluminium alloy coins having the virtues of lightness, durability and cheapness compared with other coinage metals, is a typical 'coining' operation.

Forming

Aluminium and its alloys are among the most readily formable of the commonly fabricated construction metals.

Aluminium alloys vary widely in their formability characteristics depending upon the alloy composition and its temper. The choice of an alloy for a particular application therefore depends upon the severity of the forming operation involved and on other considerations such as required strength, corrosion-resistance and surface finish.

Most of the equipment used in the forming of steel and other metals is suitable for use with aluminium alloys. However, the press force required is usually lower than for comparable operations on steel and higher press operating speeds are obtainable. Similarly equipment for roll forming, spinning, stretching and other operations need not be as massive as for comparable steel forming.

Although tool wear generated when forming aluminium is less than that with steel so that in principle tools can be made from less expensive materials, it is usual for many forming operations to use hardened steel tools. This is because these tools withstand aluminium's highly abrasive oxide surface. Surface finish of the tooling is important too, and a highly polished surface is essential to avoid marking the aluminium.

The most important point for a designer and specifier is that aluminium responds well to forming and machining operations, offering considerable design freedom whether that be expressed in flowing curves of polished extrusions for handrailing or the profiling of sheet cladding for a commercial building.

Multiple punching of a glass-house member

All of the standard metal-working practices such as sawing, cutting, punching, mitring, drilling and shearing are performed on aluminium materials, although the ground rules for dealing with a softer metal than steel are rather different – clearances, bend radii, lubrication, cutting and rake angles and saw-teeth density, for example, all require to be modified to suit the particular aluminium alloy being fabricated. Standard reference books are available that give detailed information. (The Aluminium Federation library is a good source.)

Blanking

Blanking of aluminium sheet is generally carried out on punch presses; these give a high production rate and enable the maintenance of close tolerances. Press brakes may also be used, generally for short-run production or experimental work.

Roll forming

Aluminium alloys are readily shaped by roll-forming. High operating speeds and efficient production are attainable.

The process is one well known to the architectural profession. Many thousands of tonnes of aluminium sheet have been processed into profiled

Forming profiled sheet from pre-painted aluminium coil

forms ranging from the simple sinusoidal corrugated sheeting to special designs of ridged and troughed panels.

By employing a series of cylindrical dies in 'male' and 'female' sets, sheet and plate can be progressively formed to final shape in a continuous high-speed operation.

Stretch forming

Almost all of the aluminium sheet alloys can be shaped by stretch forming. In this process the workpiece is stretched over a shaped form and stressed beyond its yield point to produce the desired contour. Compound curves, dimensional accuracy, wrinkle-free shapes and highly cost-effective components can be produced with this method. The use of super plastic aluminium alloy has extended the scope of this technique considerably enabling severely but attractively formed panels to be produced. Extrusions, too, can be shaped with stretch forming equipment.

Pressing

High-speed presses are used to make aluminium products from foil containers for takeaway meals to saucepans and TV reception dishes. The metal, normally in the soft, or annealed condition, unless only shallow

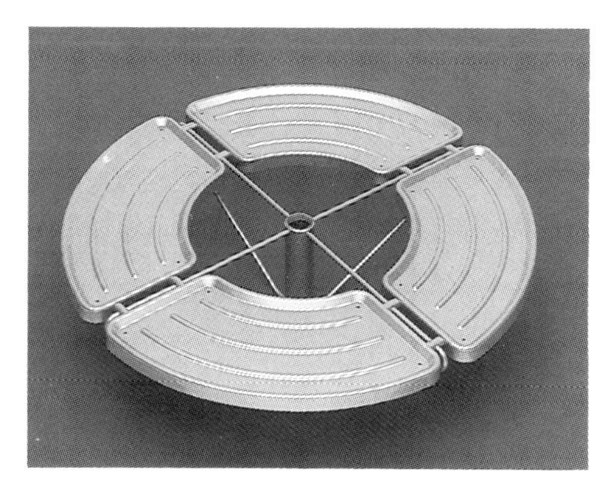

An attractive quadrant design of shop shelving made using super-forming aluminium sheet finished in a silver metallic paint and clear lacquer. The shelves were installed in the 'Way-In' department at Harrods, London

forming is involved, is formed between male and female dies with tolerances such that controlled metal movement is allowed during the forming operation.

Drawing

Seamless, smooth-sided, cup-like shapes can be produced on either single- or double-action presses depending upon the depth of draw. Electric lamp caps, fish boxes, bottle tops and drinks cans are all examples of drawn aluminium. The process for making drinks cans, now highly automated and operating with multiple-strike tooling at very high speeds, is a sophisticated multi-stage one involving blanking, drawing and then wall-ironing to elongate and thin down the cylindrical sides of the cans.

Spinning

Aluminium is an ideal material to be worked by spinning, one of the oldest metalworking crafts known. In its simplest form, a hardwood mandrel, having the required contours of the bowl-shaped product to be made, is fixed in the chuck of a spinning lathe. A flat, circular aluminium blank (known as a circle) is held firmly against the mandrel and rotated with it. By exerting pressure on the blank with a hardwood or metal former, the blank is slowly forced to take up the shape of the mandrel, while the metal thickness is reduced by up to 30 per cent.

The process has been widely used for years to produce hollow-ware but it is also used to produce much larger items such as domed ends for transport tankers and chemical plant vessels.

Lighting reflectors made from spun and anodized aluminium alloy sheet

Forming a heavy-duty architectural flashing using a computer-controlled press brake

Bending

Aluminium in sheet and some plate thicknesses may be bent on standard press brakes. Techniques are similar to those for steel, differing only in some details of tool design. In particular, bend radii for 90° cold bends are generally greater than for steel (Table 3.3). Not only sheet but tube and variously-shaped extrusions can be bent and curved.

The stresses and the metal flow in a section, particularly a hollow one, being bent are complex. The material on the outside of a bend must be permanently stretched, while that on the inside must be compressed. For tubes and pipes the tension in the metal on the outside of the bend will tend to cause flattening. Such flattening will decrease structural strength and reduce the internal diameter, causing a constriction of flow if the tube is to carry a liquid.

Thus bending conditions must provide for even stretching on the outside of the bend, control of wrinkling and buckling on the inside and internal support to prevent flattening of the diameter.

The preferred type of tube bender is the 'draw-bender', in which the point of bending remains fixed and the tube is clamped to a rotating form and drawn through the point of bending. This type of bender enables a smooth contour to be maintained during the bending operation, while a mandrel inserted in the bore minimizes any flattening.

The minimum bend radius for a particular tube size will depend on the alloy and temper of the metal and on the bending equipment. Certain basic facts influence the acceptable severity of a bend. The heavier the wall thickness the less the tendency to flatten and wrinkle. The smaller the tube diameter the easier the bending becomes because there is reduced metal flow.

The bending of solid sections follows similar principles, but it is sometimes possible to facilitate section bending by notching or cutting out metal in an area of bend compression. This removes completely the metal that would otherwise be wrinkled and buckled in the bending operation. Depending upon the particular application of the section it may or may not be desirable to weld the cut section after bending.

Furniture and window frame sections are examples of successful bending on a regular production basis, and many industries are successfully taking advantage of the scope offered by modern sophisticated bending and stretch-forming machinery.

Machining

Aluminium alloys are readily machinable and may be cut at high speeds. The metal, however, differs in its characteristics from steel, and in particular two important considerations have to be allowed for in all machining operations. Aluminium has a high coefficient of friction with steel and hence polished tools and good lubrication are essential to maintain the cutting edges and prevent tearing. Also aluminium has a higher thermal coefficient of expansion than steel and a coolant in the form of a cutting

Table 3.3 Approximate bend radii* for 90° cold bend in various aluminium alloys of different thicknesses and tempers

Alloy	Temper	Radii for various thicknesses (in mm) expressed in terms of thickness 't'							
		0.40	0.80	1.6	3.2	4.8	6.4	9.5	12.7
1200	O	0	0	0	0	½t	1t	1t	1½t
	H2	0	0	0	½t	1t	1t	1½t	2t
	H4	0	0	0	1t	1t	1½t	2t	2½t
	H6	0	½t	1t	1½t	1½t	2½t	3t	4t
	H8	1t	1t	1½t	2½t	3t	3½t	4t	4½t
2014A	O	0	0	0	½t	1t	1t	2½t	4t
	T4	1½t	2½t	3t	4t	5t	5t	6t	7t
	T6	3t	4t	4t	5t	6t	8t	8½t	9½t
3103	O	0	0	0	0	½t	1t	1t	1½t
	H2	0	0	0	½t	1t	1t	1½t	2t
	H4	0	0	0	1t	1t	1½t	2t	2½t
	H6	½t	1t	1t	1½t	2½t	3t	3½t	4t
	H8	1t	1½t	2t	2½t	3½t	4½t	5½t	6½t
5454	O	0	½t	1t	1t	1t	1½t	1½t	2t
	H2	½t	½t	1t	2t	2t	2½t	3t	4t
	H4	½t	1t	1½t	2t	2½t	3t	3½t	4t
6082	O	0	0	0	1t	1t	1t	1½t	2t
	T4	0	0	1t	1½t	2½t	3t	3½t	4t
	T6	1t	1t	1½t	2½t	3t	3½t	4½t	5t
7075	O	0	0	1t	1t	1½t	2½t	3½t	4t
	T6	3t	4t	5t	6t	6t	8t	9t	9½t

* The radii listed are the minimum recommended for bending sheets and plates without fracturing in a standard press brake with air bend dies. Other types of bending operations may require larger radii or permit smaller radii. The minimum permissible radii will also vary with the design and condition of the tooling.

Curving a length of specially profiled fascia trim for a Ford Motor Company forecourt showroom

liquid is generally necessary to prevent excessive stresses in the work or the machine.

Chip-forming characteristics also vary from alloy to alloy. Lower-strength wrought alloys produce long, ribbon-like chips whereas the high-strength heat-treatable alloys produce relatively short chips. For the large-scale production of repetition machined-parts a special machining alloy containing small percentages of lead and bismuth offers the machinist characteristics similar to those of free-machining brass. Aluminium casting alloys have good machining characteristics particularly those containing copper or magnesium.

In general it may be concluded that, given the correct choice of tooling material, angles, lubrication and cutting speed, aluminium alloys in all forms from thick aircraft alloy plate to small parts machined from bar, and from automotive castings to milled, drilled and tapped furniture fittings may be machined with ease.

The reception area at Laporte Industries offices, Luton, gets a bright, clean appearance with this use of aluminium fenestration involving both rectangular and curved windows

Computer-programmed machining of an aluminium extrusion forming part of a high-tech computer station furniture unit (courtesy Shape Engineering)

4 Joining

For many applications it is necessary to use a joining process, and there is a wide range of methods available for aluminium, including soldering, brazing, welding and adhesive bonding. These processes are all widely used commercially to join products ranging from aircraft skins and chemical tankers to pan handles and transformer windings.

It is not all that long ago since aluminium was regarded as difficult to join other than by mechanical methods. This was because the processes applied were those already in use for steel, a much older and more established metal, and they were not fully suited to aluminium, with its different metallurgical characteristics.

Today, processes are available that enable wrought aluminium to be joined whatever its thickness or alloy from the thinnest of electrical foils to the thickest piece of armoured plating.

Characteristics

There are a number of particular characteristics of aluminium to bear in mind when considering joining – particularly soldering, brazing and welding.

1 All aluminium alloys have a tenacious, hard, oxide film that forms instantly when an aluminium surface is exposed to air. This oxide has a very high melting point (around 2000°C), and is virtually insoluble

in the metal; thus inhibiting wetting by molten filler metals. This oxide skin therefore needs to be removed or broken up during the welding process to ensure good wetting and fusion.

2 Aluminium has thermal and electrical conductivities approximately four times greater than steel. Higher heat inputs are therefore necessary for fusion welding with aluminium, and also increased current for resistance welding.
3 The metal is highly reflective to radiant energy including visible light and heat. Unlike steel it does not assume a colour change as the fusion temperature is approached and so judgement criteria as to whether the metal is becoming molten are changed.
4 Aluminium is non-magnetic and so 'arc blow' is not encountered with aluminium as it can be with steel.
5 The linear coefficient of thermal expansion is about twice that of steel. However, the temperature melting range is only about half. Thus during welding, the expansion for identical aluminium and steel components is about equal as the effects of the expansion coefficient and melting range tend to balance each other out.
6 The heating applied during soldering, brazing or welding has a softening effect on the metal in the region of the join. This means, for example, that work-hardened alloy material will be locally annealed. This needs to be allowed for in the component design, the selection of the joining process (Table 4.1) and the manufacturing procedures.

Alloys

The alloying of aluminium by the addition of quantities of other elements such as silicon, copper, magnesium, zinc and lithium produces alloys with a wide range of differing properties. These changes in properties determined by the metallurgical composition selected are accompanied by changes in joining characteristics. Some alloys respond much better than others to different joining processes (Table 4.2). However adhesive bonding, subject to the choice of correct metal surface pre-treatment, and also ultrasonic and solid-state welding methods, are effective on all alloys. The latter two welding methods have only specialized applications and are not in general usage.

Welding

The two most important and widely used methods for the fusion welding of aluminium are the Tungsten Inert Gas (TIG) and Metal Inert Gas

Table 4.1 Principal joining processes

The most commonly used processes	
Fusion welding	TIG
	MIG
	electron beam
	laser beam
	oxyacetylene (possible but not recommended for good quality work)
	electrogas
	electroslag
	submerged arc
Soldering	
Brazing	torch
	flux-dip
	vacuum/inert gas
Adhesive bonding	
Resistance welding	spot
	seam
Stud	
Flash-butt	
Other less commonly used processes include:	
Solid state	friction
	explosive
	ultrasonic
	cold pressure
	hot pressure

(MIG) processes.

With these two processes, fluxes are not required because the use of an inert gas shield prevents re-oxidation of the metal surface after the cleansing action of the electric arc has removed the oxide.

Both these processes give high-quality welds in stressed structures and are suitable for welding in all positions. It is possible, depending upon the application and chosen process, to weld thicknesses from 0.5 mm up to 75 mm or more. This thickness range is extended by the use of automatic procedures. Development work within the industry is seeking to extend the maximum thickness for welding. This is of particular importance for off-shore oil rig applications where increased use of aluminium is anticipated for above sea-level structures, including accommodation and housing units.

Table 4.2 Joining suitability

Alloy	TIG/MIG	Resistance	Brazing	Soldering
1080A	V	G	V	V
1050A	V	V	V	V
1200	V	V	V	V
2014A	N	E	N	N
3103	V	E	V	V
3105	V	V	G	G
5005	E	E	G	G
5083	E	E	N	N
5154A	E	E	N	N
5251	V	E	N	N
5454	E	E	N	N
6061	V	V	V	G
6063	V	V	V	G
6082	V	V	G	G
7010	V	V	N	N
7020	V	V	N	N
7075	N	V	N	N

E = Excellent V = Very good G = Good N = Not recommended

TIG

In this process an alternating current arc is struck between a tungsten electrode and the aluminium workpiece. A shroud of inert gas covers the electrode and the weld area. A filler rod, if required, is fed in independently.

The method allows close control by the operator of both the heat input and the amount of filler material fed into the weld and it is particularly applicable for the intricate torch manipulation required for pipework and complex structures.

MIG

In the MIG process a direct current arc of reverse polarity (with the electrode positive), shielded by an inert gas shroud, is struck between the aluminium and a continuously fed aluminium wire electrode which undergoes controlled melting at the tip and so acts as the filler material.

The process lends itself to high-speed automatic welding but lacks the penetration control possible with TIG welding.

Filler wires

A variety of filler wires in different alloys and diameters is available. It is important to select the appropriate filler for the alloy being welded in

Fabricating an aluminium sheet architectural corner piece using TIG manual welding

order to achieve compatibility, and particularly to avoid the possibility of weld cracking (Table 4.3).

Brazing

Many aluminium alloys can be brazed. The process is widely used commercially and has particular importance nowadays for the production of automotive radiators.

The filler metals used are invariably aluminium–silicon alloys with a silicon content ranging from about 7.5 per cent up to 12 per cent. For many sheet products an aluminium sheet clad with a thin layer of Al–Si alloy sheet is frequently used, particularly for furnace or dip brazing, thus avoiding the need for an applied filler metal. Such material, known as brazing sheet, can be formed easily by conventional means.

As with soldering, chemical fluxes are required except where furnace brazing is carried out in vacuum or inert gas. Any corrosive flux residues must be removed after brazing, and after proper cleaning brazed joints have excellent corrosion resistance. Much development work has been done on non-corrosive fluxes and Nocolok is one proprietary flux that gives good results.

Table 4.3 Selection of filler rods and wires for MIG and TIG welding

Parent Metal Combination	7020	6082	6063	6061	6101A	5083	5454	5154A	5251	5005	3105	3103	1200	1050A	1080A
1080A and 1050A	5556A 5556A 5556A 5556A	4043A 4043A 4043A 5056A	4043A 4043A 4043A 5056A	4043A 4043A 4043A 5056A	4043A 4043A 4043A NA	5556A 5556A 5556A 5556A	5056A 5154A 5056A 5056A	5056A 5056A 5056A 5056A	5056A 5154A 5056A 5056A	5056A 5056A 5056A 5056A	4043A 3103 4043A 1050A	4043A 3103 4043A 1050A	4043A 1050A 4043A 1050A	4043A 1050A 4043A 1050A	4043A 1080A 4043A 1080A
1200	5556A 5556A 5556A 5556A	4043A 4043A 4043A 5056A	4043A 4043A 4043A 5056A	4043A 4043A 4043A 5056A	4043A 4043A 4043A NA	5556A 5556A 5556A 5556A	5056A 5154A 5056A 5056A	5056A 5056A 5056A 5056A	5056A 5154A 5056A 5056A	5056A 5056A 4043A 5056A	4043A 3103 4043A 1050A	4043A 3103 4043A 1050A	4043A 1050A 4043A 1050A		
3103	5556A 5556A 5556A 5556A	4043A 4043A 4043A 5056A	4043A 4043A 4043A 5056A	4043A 4043A 4043A 5056A	4043A 4043A 4043A NA	5056A 5056A 5056A 5056A	5056A 5154A 5056A 5056A	5056A 5154A 5056A 5056A	5056A 5154A 5056A 5056A	5056A 5056A 5056A 5056A	4043A 3103 4043A 3103	4043A 3103 4043A 3103			
3105	5556A 5556A 5556A 5556A	4043A 4043A 4043A 5056A	4043A 4043A 4043A 5056A	4043A 4043A 4043A 5056A	4043A 4043A 4043A NA	5056A 5056A 5056A 5056A	5056A 5154A 5056A 5056A	5056A 5154A 5056A 5056A	5056A 5154A 5056A 5056A	5056A 5056A 5056A 5056A	4043A 3103 4043A 3103				
5005	5556A 5556A 5556A 5556A	4043A 4043A 4043A 5056A	4043A 4043A 4043A 5056A	4043A 4043A 4043A 5056A	4043A 4043A 4043A NA	5056A 5056A 5056A 5056A	5056A 5154A 5056A 5056A	5056A 5154A 5056A 5056A	5056A 5154A 5056A 5056A	5154A 5154A 5056A 5154A					
5251	5556A 5556A 5556A 5556A	5056A 5056A 5056A 5056A	5056A 5056A 5056A 5056A	5056A 5056A 5056A 5056A	5056A 5056A 5056A NA	5056A 5056A 5056A 5056A	5056A 5154A 5056A 5056A	5056A 5154A 5056A 5056A	5556A 5554 5056A 5056A						
5154A	5056A 5556A 5056A 5056A 5056A	5056A 5056A 5056A 5056A	5056A 5056A 5056A 5056A	5056A 5056A 5056A 5056A	5056A 5056A 5056A NA	5056A 5556A 5054A 5056A 5056A 5154A	5056A 5554 5056A 5154A	5056A 5154A 5056A 5154A							
5454	5056A 5556A 5056A 5056A 5056A	5056A 5056A 5056A 5056A	5056A 5056A 5056A 5056A	5056A 5056A 5056A 5056A	5056A 5056A 5056A NA	5056A 5556A 5154A 5056A 5056A 5154A	5056A 5554 5056A 5554								
5083	5556A 5556A 5556A 5056A	5556A 5556A 5056A 5556A 5056A	5556A 5556A 5056A 5556A 5056A	5556A 5556A 5056A 5556A 5056A	5556A 5556A 5056A 5556A NA	5556A 5556A 5556A 5056A									
6061	NR NR NR NR	4043A 4043A 4043A 5056A	4043A 4043A 4043A 5056A	4043A 4043A 4043A 5056A											
6082	NR NR NR NR	4043A 4043A 4043A 5056A													
7020	5556A 5556A 5556A 5556A														

Recommended filler materials for each combination of materials to be welded are shown in one box, which is located by traversing horizontally and vertically from the appropriate parent metals. The filler material for maximum weld strength is shown on the top line. The filler material for maximum resistance to corrosion (see CP 118 and BS 1500 Part 3) is shown on the second line. The filler material for freedom from persistent weld cracking is shown on the third line. The filler material for optimum colour matching when welds are to be anodized is shown on the bottom line. Note: NA = Not applicable. NR = Parent metal combinations not recommended.

Brazing joint strength is very high and joints can be made rapidly and inexpensively. Joints that are inaccessible and so not joinable by other methods can often be joined by brazing. Automotive radiators are a good example.

Soldering

Removal of the aluminium oxide skin is the important part of all aluminium soldering operations. This removal operation generally requires the use of an active corrosive flux, although fluxless soldering using a mechanical means of abrading the oxide skin is effective for certain joint configurations. Fluxes for aluminium are inherently corrosive and any residue left after soldering can cause corrosion problems, and must therefore be removed.

Various solders have been developed for use with aluminium alloys. These can be classified into three groups – low, intermediate and high temperature – depending upon the constituents of the solder. In general it may be stated that the joint strength and corrosion resistance increase as the soldering temperature increases, but conversely the ease of application decreases.

Most aluminium alloys can be soldered.

Joint preparation

With all fusion welding, brazing and soldering, it is essential to thoroughly clean the joint area prior to welding. Additionally, depending upon the metal thickness, edge preparation (such as tapering for example) may need to be carried out.

Adhesive bonding

Adhesive bonding of aluminium has been carried out successfully for many years, particularly in the aircraft industry. Recently it has assumed increasing importance in the automotive industry where its use is enabling very stiff aluminium structures to be assembled. Increased interest in adhesive bonding is being shown in the general engineering industry.

A very high strength-to-weight ratio is achievable with this method.

Laser welding

Laser welding is still very much in its infancy as a commercial method for joining aluminium, but it is being used very successfully to make seam-welded tube for double-glazing spacer bars.

The method suffers from the inherent problem caused by the high reflectivity of aluminium, which dissipates much of the incident light energy striking the metal surface during welding.

Joining to other metals

Various techniques exist for joining aluminium to other metals, with adhesive bonding being perhaps the most adaptable (and also usable for aluminium-to-plastics joints). Solid-state welding methods such as friction, explosive and cold pressure are all valuable for joining dissimilar metals.

5 Finishing

Introduction

One of the advantages of aluminium is that it is a naturally durable metal which can be used in many applications without the need for added protection against the ravages of atmospheric or other chemical attack. Profiled building sheet, greenhouse frames, ladders, and lorry bodies are examples of uses where the natural 'mill' finish of aluminium is perfectly acceptable.

Many products, however, call either for added protection or a decorative appearance or even a combination of the two. For such applications there are a number of suitable options using aluminium.

The metal may be:

- mechanically treated to produce a variety of attractive textures;
- chemically treated to produce a degree of atmospheric protection;
- anodized;
- coated;
- enamelled;
- chrome-plated;
- lacquered;
- printed.

Of these, anodizing and coating are in widespread use across industrial and architectural applications, with lacquering and printing being used extensively in canning and packaging applications of aluminium.

Anodizing

Immediately aluminium and aluminium alloys are exposed to oxygen a very thin layer of aluminium oxide forms on the metal's surface. This oxide film, about 0.01μm thick, provides a resistance to atmospheric corrosion and it is the presence of this thin protective oxide coating that gives aluminium its natural durability.

Anodizing, an electrolytic process that is commercially unique to aluminium, enables the oxide film to be thickened by a factor of more than one thousand. The resultant anodic coatings have improved physical and chemical properties together with colouring possibilities that have greatly increased the range of applications for aluminium and its alloys. Widely used and accepted in the building industry as the outstanding finishing method for aluminium, anodizing has, over a number of years, opened up diverse engineering applications. These range from nameplates and tea trollies where decorative aspects predominate, to engine parts where abrasion-resistance coatings are an important requirement.

The US Embassy in Grosvenor Square, London was fitted with gold anodized aluminium windows in 1959. The 40 foot golden eagle that dominates the frontage facing the square was constructed of cast and anodized aluminium components

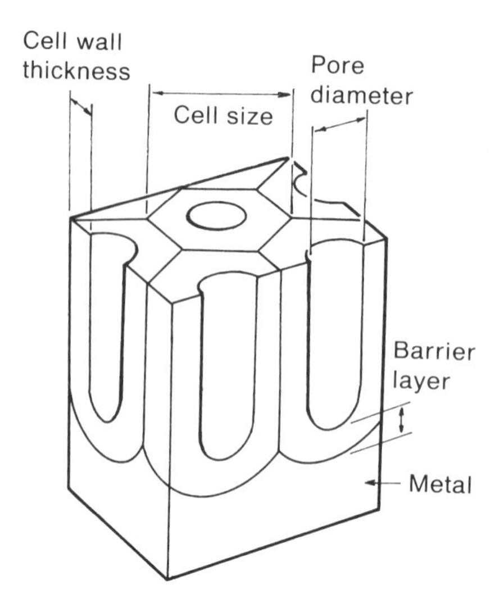

Figure 5.1 Microstructure of anodic film on aluminium

Principles of anodizing

When a piece of aluminium is the anode in an electrolytic cell and a current is passed through, oxygen, instead of being liberated at the anode as with some metals, reacts with the aluminium to form a layer of porous aluminium oxide. The amount of aluminium oxide formed is directly proportional to the current density and time. The structure of the oxide layer comprises microscopic hexagonal columns each with a central pore.

The diameter of the pores and the thickness of the barrier layer that is continuously formed during the anodizing process between the underlying metal and the growing cells are controlled not only by the particular electrolyte used but also by the temperature and voltage applied. Thus by varying the anodizing conditions and the electrolyte itself, it is possible to alter the physical properties of the coating, such as its hardness, abrasion resistance, density and colour characteristics. It is this flexibility that enables anodized aluminium to have so many different applications. Figure 5.1 shows the anodic film structure on aluminium.

Basic anodizing processes

The properties of anodized aluminium depend upon a combination of the following factors:

- the aluminium alloy used;

Silver anodizing of extruded lengths (courtesy Finalex)

- pre-treatment process;
- the anodizing process;
- post-anodizing processes.

Many chemical solutions have been used, or proposed, for anodizing electrolytes; most of these are acidic but some alkaline solutions have been considered. The most commonly used electrolytes are those based on sulphuric acid, but other acids such as oxalic, chromic and phosphoric acid are operated on a successful commercial basis where special types of coatings are required.

Sulphuric acid Processes based on sulphuric acid are those that are generally preferred for both decorative and protective applications. On all except a few alloys containing insoluble constituents, such anodizing provides semi-transparent colourless films in thicknesses up to 35 μm. The appearance of these coatings is significantly affected by the original underlying surface finish of the aluminium, and so attractive textures such as linishing and scratch brushing are frequently applied to the metal prior to anodizing. 'Bright' anodizing, which gives a highly reflective appearance, is achieved by selecting an alloy based on high-purity aluminium and by subjecting the metal to mechanical polishing and chemical brightening prior to the electrolysis.

Bronze anodized curtain walling

Bronze anodized fascia extrusions and window frames are used to provide a striking appearance at Chester Northgate

A process variation of considerable engineering importance is the use of low-temperature anodizing. Normally sulphuric acid bath temperatures range between 18° and 25°C but by operating at temperatures between −5° and +5°C very hard coatings are obtained. 'Hard anodizing' is finding increased applications in engineering where abrasion resistance is of critical importance, such as in moving machine parts.

Chromic acid Chromic acid was used in the first commercial anodizing process invented in 1923 by Bengough and Stuart. The process produces thin films that are usually opaque grey in colour. Knitting needles are a familiar but declining example. The process is still widely specified, and is particularly used in the treatment of aircraft components.

Oxalic acid Solutions of this acid tend to produce translucent, hard, yellowish coatings that in the past have had some acceptance for architectural applications where an 'integral colour' obtained without the use of added dyes or pigments has been required. The coatings have a higher abrasion resistance than those obtained by sulphuric acid, but the higher process costs and limited colour appeal have restricted the use of this acid.

Phosphoric Phosphoric acid produces a coating structure with larger pore diameters than in conventional sulphuric acid anodizing. The structure is ideally suited as a pre-treatment for the electro-plating of aluminium, and as a pre-treatment for the adhesive bonding of aluminium components.

Coloured anodic films

Anodic films can be produced in colour by a variety of methods. Many of the colours are extremely light-fast, and so find widespread appeal in exterior architectural applications. Colour anodizing presents a lustrous visual quality that is much appreciated by architects and their clients.

Colour is achieved in three principal ways:

- absorption of added colouring agent by the freshly produced anodic film;
- integral colouring;
- electrolytic two-stage coloration using a metal salt.

Absorption A freshly produced anodic film made by the sulphuric acid process has a porous absorptive structure that readily accepts dyestuffs. Where no colouring is required the coating is sealed, generally in steam or boiling water, but where colour is called for, the material is immersed in an aqueous medium containing either an organic dye or an organic pigment, prior to the sealing operation.

Dyes and pigments have different degrees of light fastness, and some are more suitable than others for outdoor exposure. For many years black has been a highly successful architectural finish and in recent years other colours – yellow, gold, dark blue, turquoise blue and red – have been added.

Integral colouring The oxalic acid process was one of the earliest 'integral' colour processes – a one-stage anodizing operation in which the combination of choice of alloy and choice of electrolyte results in a naturally developed colour within the built-up oxide skin. The oxalic acid process produces yellowish films. Integral colouring using sulphuric acid anodizing and specially developed alloys has been developed for architectural applications with a range of colours from pale gold through to bronze and black. Colours produced in this way are extremely durable and light-fast, and integral colouring has built up an outstanding record of fastness performance. The coatings are harder than those of conventional sulphuric acid anodizing and have good abrasion resistance. They are frequently referred to as 'hard' coatings, but should not be confused with the low-temperature hard-anodizing method described earlier.

Electrolytic The principles of electrolytic colouring go back to the 1930s, but it was subsequent research by Professor Asada of Tokyo University that led to the world-wide commercial exploitation of this process. This two-stage process consists basically of, first, anodizing a piece of aluminium by the conventional dc sulphuric acid method and then, second, submitting the freshly anodized material to a second electrolytic process using an ac current (generally) in a metal salt solution. Many metallic salts can be used but the most reliable and controllable processes use either nickel, tin or cobalt.

The process deposits metal at the base of the anodic pores and the amount of deposition determines the perceived colour shade, which varies from pale bronze to black. These colours are the result of light wavelength scatter and reflection in the anodic film and are not dependent upon colouring additives that could fade under exposure to ultra-violet light. The light fastness of the coatings is therefore excellent.

Combination colouring Where colour fastness is not a long-term necessity, then the colour range available for anodizing aluminium is very wide, and there are few restrictions on colour choice. For architectural applications, where light fastness for many years is required, the range of colours has, in the past, been mainly restricted to greys, bronzes and black. Recent commercial development work on colour finishing by applying two different colouring systems in succession has significantly widened the choice of durable colours now available. Combination finishing expands the scope for colour styling in architecture and retains the aesthetic lustrous quality of the metal.

In essence, combination colouring consists of first producing a coloured effect either by an integral or an electrolytic method and then impregnating the anodic film with a colour pigment. Using the electrolytic process, plus additional colour pigment enables a highly attractive range of durable gold, red and blue colours to be obtained.

Applications of anodized aluminium

Anodizing provides both protection and decoration for aluminium in all forms – cast, rolled and extruded. Many engineering and military applications use anodizing purely as a means of protection against corrosion, for example shell cases, rocket components, fuses and gun barrels. Other engineering uses are increasingly exploiting the combined advantages of using aluminium for its lightness, strength and, particularly with extrusions, for the shape potential. The latter frequently leads to easier component production and assembly, plus the decorative appeal of a colour anodized finish.

Special anodizing quality sheet is used for the profiled fascia that forms a distinctive part of the current Ford Motor Company corporate identity livery on its showrooms world-wide

Medical and scientific instruments, electronic components and cabinets, shower cubicles and lighting units are just a few examples of diverse applications of anodized aluminium. In the architectural and building area, colour anodizing is being widely specified, and an interesting example that typifies the properties and benefits of anodized aluminium is the new style payphone booths that are now installed throughout the UK. These have integral colour black anodized aluminium alloy in their construction in order to provide both a durable and scratchproof anti-graffiti surface.

Coating thickness

Except in one or two specialized instances it is unusual to produce anodic coatings above 25 μm. In the building industry the thicknesses used range from 5 μm up to 25 μm depending upon the application. Exterior applications where long-term durability is required call for thicknesses of 25 μm. For recommended thicknesses refer to BS1615/1987 'Specification for Anodic oxidation coatings on aluminium', and BS3987/1974 'Specification for Anodic oxide coatings on wrought aluminium for external architectural applications'.

Testing the film thickness of anodized sections

Paint and lacquer coatings

Coatings for aluminium are available in a range of compositions and can be applied by a number of different production techniques. These coatings are available in a wide spectrum of colours that will retain their appearance for many years and will protect completely the underlying metal. This combination of colour and protection, coupled with the ability of today's coating specialists to apply finishes to consistently high, industry-approved standards has resulted in coloured aluminium being used more and more widely. Extruded sections for domestic windows are generally coated in white, but other colours, particularly brown, are applied, while sections for commercial fenestration and cladding are employed in a myriad of colours to suit the specifier's requirements.

Sheet, too, is finished with a variety of colour coatings. Nowadays there is very little 'mill finish' profiled sheeting used on buildings, although its use from the 1950s onwards for many applications demonstrated the intrinsic durability and functionality of aluminium sheet as a building material.

Coating compositions have been developed by the paint industry to satisfy a wide range of requirements from surface protection to decoration. Compositions include liquid coatings based on acrylic, pvc, polyester, polyurethane and fluoropolymer (such as polyvinylidene fluoride) resins,

Coils of pre-painted aluminium in readiness for roll-forming into profiled building sheet

Jigging aluminium extrusions ready for powder coating (courtesy Hydro Aluminium Century)

Table 5.1 Typical coatings and their properties (applied to pre-painted coil)

Type of coating	Polyester	PVF	PE/PU/PA	Alkydamine	PVC Plastisol	Silicone polyester
Primer	None	Epoxy	None	None	Epoxy	Epoxy
Film thickness (Nom)	20/25 μm	28 μm	30/35 μm	20/25 μm	100/200 μm	25 μm
Gloss (60°)	8–90+	30	30	8–90+	–	30%
Hardness	H	F–H	2H	F–H	–	F–H
Bend test	2T–1T	1½T	1T	2T	1T	5T
Appearance	Smooth	Smooth	Slight texture	Smooth	Textured	Smooth
Colour restrictions	None	Bright colours	None	None	Limited range	Limited range
Metallic colours	Yes with life restrictions	Yes with life restrictions	Yes	Yes with life restrictions	No	Yes with life restrictions
Handleability	Fair	Fair	Excellent	Fair	Excellent	Fair
Abrasion resistance	Fair	Fair	Excellent	Fair	Excellent	Fair
Chemical resistance	Good	Excellent	Excellent	Fair	Fair	Fair
UV resistance	Good	Excellent	Excellent	Poor	Fair	Good
Temperature resistance	120°C	120°C	140°C	120°C	60°C	120°C
Ease of cleaning	Good	Very good	Very good	Good	Poor	Very poor

and powder coatings based on polyester, acrylic and polyurethane resins. Table 5.1 shows some typical coatings and their properties.

Aluminium can be coated prior to forming into components, or as assembled fabrications. For both of these basic types there are now established coating specialists that carry out in-house rather than on-site painting. This approach brings uniformity of finishing and a high standard of quality control and specification.

The main coating methods are electrostatic spray, used for both liquid and powder applications, various immersion methods, roller and curtain coating. As well as hand spray processes, mainly used on finishing components, highly sophisticated automatic flow lines for extrusions, sheet and coil are employed.

For all of these types of coating correct preparation of the metal surface is essential to ensure good adhesion and in-service performance of the coating. However good the paint formulation the end result can be ruined by inadequate attention to pre-treatment. Pre-treatment is one of the

Keeping a watchful eye on the electrostatic powder coating of aluminium extrusions (courtesy Hydro Aluminium Century)

features that distinguishes a factory-applied finish from a site-applied finish.

The generally adopted way of applying an organic coating to aluminium sheet and extrusions is by electrostatic spray. In this process the pretreated items are sprayed either manually or automatically, and subsequently stoved at a temperature of around 220°C for about 30 minutes. The coating heads are charged to a voltage of up to 70 kV. The electrostatic nature of the process ensures that the coating gets into all of the grooves, channels, slots and drilled holes in the extrusions.

Powder coatings have become increasingly popular, particularly since the oil crisis of 1973. At that time solvent costs started to rise rapidly thus making powder coatings financially more attractive than liquids. Environmentally too, powder coatings are now preferred because of the avoidance of solvent dispersed into the atmosphere, which can occur with liquid coating methods unless very stringent process control is observed in production.

Coatings based on polyester resins are widely available and now dominate over other powder resins. These polyester coatings have weathering characteristics comparable to the liquid acrylic coatings, and have very good mechanical performance. Powder coatings can be applied to greater thicknesses in a one-coat operation than can liquid coatings and down to a minimum thickness of between 50 and 60 μm.

A garden centre roofed in profiled, pre-painted aluminium coil with distinctive triangular roofing

Thorough pre-treatment with as many as ten cleansing operations, ensures that these coatings can withstand a whole series of mechanical abuse tests. The tests include conical bends, sawing, scratching, gouging, hammering and weight drops. Both liquid and powder coatings provide strongly adherent weather-resistant coatings in attractive colours.

Electrophoretic coating is a refinement of the simple paint dipping process. In some ways the process is akin to anodizing, because the metal to be coated is made the anode in an electrolytic bath. In this case it is a water-dispersable paint that is the electrolyte. A direct current is passed through the bath and paint solids are deposited on the aluminium.

The process is ideal for a continuous automatic line, as well as a batch process. A typical sequence of operations begins with cut-to-length sections being sprayed with alkaline chemicals and then passed through a series of rinses ending with a rinse in de-ionized water, after which the sections, now thoroughly clean, are ready for coating. The sections are then dipped in the paint, at the same time becoming the positive electrode of an electrolytic cell. A current is passed through the paint, and paint solids are deposited on the metal. After deposition, excess paint is rinsed off and recycled back into the main coating tank. A final water rinse is performed before the sections proceed to the stoving oven. Curing at a temperature of around 200°C for about 30 minutes completes the operations, following which the sections are ready for assembly.

Coatings applied in this manner have a very smooth surface finish, almost entirely free of the slight orange-peel effect that can be detected with other methods. Because of the highly automated nature of the process, it is only suitable for long continuous runs so, at the time of writing, only white or bronze colours are available, as demand for other individual colours is insufficient to justify using this method.

The electrophoretic process provides good penetration of paint into all recesses and corners of the sections, with a minimum thickness of 20 μm. The close control exercised throughout the operation and thorough attention to surface preparation prior to coating ensures that the coating, which can be up to 40 μm thick, adheres thoroughly. This method has however lost favour in recent times to the powder spray process.

Large quantities of aluminium coil are colour coated prior to further processing into profiled cladding or components. This highly specialized operation involves large coils of aluminium being first pre-treated and then passed through a paint-application station that applies a precise, even liquid coating across the coil web followed in turn by a curing and drying stage and then a final recoiling operation.

Care and maintenance

Both anodized and coated aluminium finishes are tough and durable. Anodizing produces a surface that is an integral part of the metal itself and this cannot flake off or chip. Applied organic coatings, whatever the application method, subject of course to following the correct procedures, do not suffer from flaking or spalling. Because of aluminium's inherent durability, these are no potential corrosion hazards resulting from the bare edges that arise from any post-coating fabrication, such as cutting and drilling. Both types of finish therefore eliminate the need for periodic attention to make up for loss of the protective coating. In this sense the coatings are 'maintenance free'. Unfortunately there has been a tendency to take this phrase too literally and to equate 'freedom from maintenance' with 'OK to neglect'. This is certainly not the case. All aluminium finishes require regular, even if infrequent, cleaning to maintain appearance and to ensure acceptable service life for the finish. Dirt, grime, and air-borne chemical pollutants will, if not removed, spoil an aluminium finish as they will any other surface or finish.

Solvents and abrasives should be avoided in any cleaning. The general rule is to clean outside surfaces with a mild detergent and water at something like three-monthly intervals depending upon the severity of the environment.

6 Durability

Aluminium is accepted as one of the most durable of metals. Some of the earliest uses of aluminium since the metal's commercial usage first started just over a hundred years ago are still in excellent condition today. Two classic examples are the statue of Eros in Piccadilly Circus, London, which was erected in 1893, and the cupola roof of the San Gioacchino church in Rome, fitted in 1897. The first example is of castings and the second of rolled sheet, with different alloys being involved in each case. Outstanding examples of longevity in architectural extrusions do not date back so far, but the window frames of the Bodleian Library, Oxford, made and installed in 1934 are still in excellent condition after well over fifty years of trouble-free service.

Since those early pioneering days the uses of aluminium in all its forms, wrought and cast, have expanded greatly, with free-world consumption running at a rate approaching 20 million tonnes annually, with the building and construction industry taking a share of about 20 per cent.

The oxide protector

It is the thin, inert layer of transparent aluminium oxide present on all surfaces of aluminium exposed to the atmosphere that gives aluminium and its alloys such good protection against corrosion. The oxide film, approx. 0.01 μm thick, is chemically stable, hard, strongly adherent to the underlying metal and has a melting point of 2000°C. This film reforms spontaneously if cut or abraded, providing that oxygen is present, and thickens slowly with age. Anodizing (see page 47) is the process of thickening this natural film in order to enhance the protection it provides to the metal beneath.

Atmospheric pollution

The atmosphere in the UK and elsewhere, depending upon whether the location is rural, coastal, suburban or industrial, can be polluted by gases such as sulphur dioxide, hydrogen chloride, ammonia, carbon monoxide and carbon dioxide. Of these, it is the sulphur dioxide and hydrogen chloride that are the most important regarding contamination of aluminium. These two gases are found in greatest concentrations in industrial areas; in urban areas the level is lower, and in rural areas the level of pollution is very low. The sodium chloride content of marine, coastal atmospheres is less aggressive to aluminium than the industrial pollutants, and so generally aluminium is regarded as a good material to use in marine environments.

Rainwater has a cleansing action on aluminium, and in the main, surfaces exposed to the elements suffer less pollution damage than more sheltered ones.

Weathering

The as-produced mill finish on wrought aluminium is naturally bright and shiny. Inevitably this brightness is dulled and tarnished by continued atmospheric exposure. The oxide darkens to a degree that is dependent upon the amount of the atmospheric pollution, its situation relative to the cleansing action of rainwater, and the amount, if any, of maintenance and cleaning applied.

In the presence of moisture, sulphur dioxide and hydrogen chloride gases form acids that can attack the metal through any weak points in the protective oxide. Corrosion pits occur at such points, and the acidic action results in the formation of corrosion by-products, which having a volume greater than the metal from which they were formed, fill and seal the pit – thus causing attack at this point virtually to cease.

This cessation of attack is a vital factor in the longevity of aluminium. It means that any attack on mill-finish, untreated aluminium, takes place in early years only and then virtually stops completely. This is illustrated by the typical corrosion time curves shown in Figure 6.1.

Aluminium windows are fitted to the top of Blackpool Tower, providing good resistance to the marine environment

Expected life

It is not possible to generalize with great accuracy on the expected life of mill-finish aluminium exposed to the atmosphere. In rural, clean atmospheres, an unlimited life may be predicted, whereas in heavily polluted areas much shorter life is inevitable. The many examples throughout the UK of aluminium in service in all kinds of atmospheres, however, make it clear that long service can be confidently expected, and examples of longevity abound. The added protection afforded by anodizing and coating ensures that the life in service of aluminium cladding, roofing, fenestration, balustrading and structures of all kinds will be very high.

Interior applications

In the interiors of domestic and office buildings, where the atmosphere is usually clean, there will be little change in the appearance of the metal apart from perhaps a slight dulling and darkening in certain areas. Generally, however, interior applications are anodized or painted to provide decorative colour.

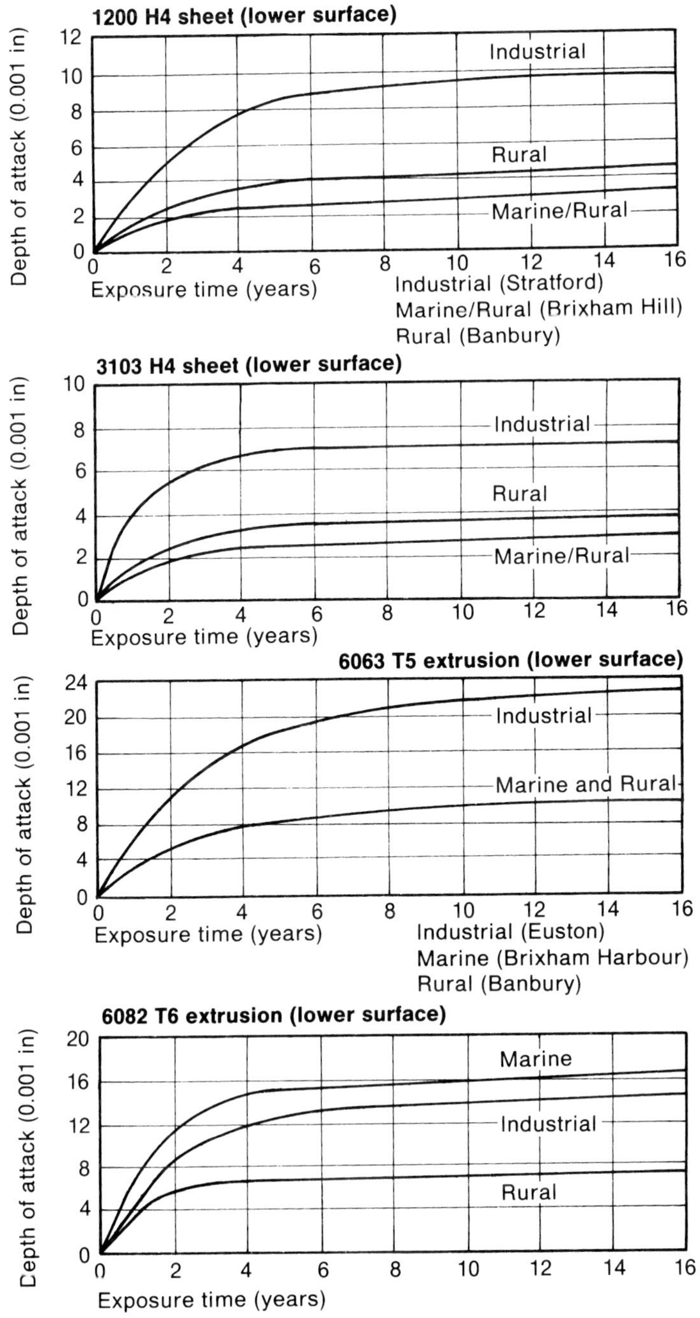

Figure 6.1 Maximum depth of attack by pitting in various environments (based on work carried out at Alcan International, Banbury)

Table 6.1 Galvanic series

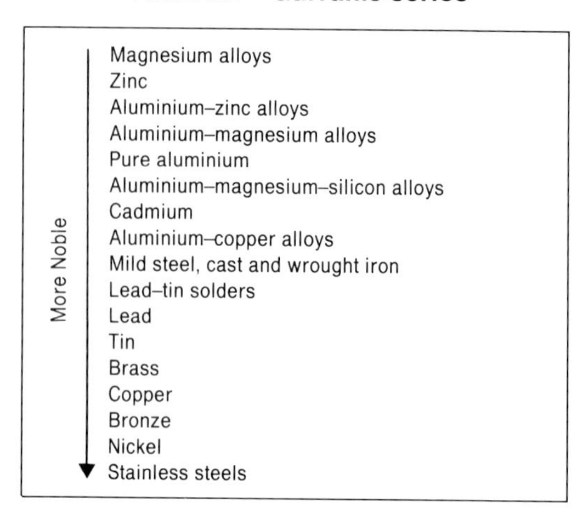

More Noble ↓
Magnesium alloys
Zinc
Aluminium–zinc alloys
Aluminium–magnesium alloys
Pure aluminium
Aluminium–magnesium–silicon alloys
Cadmium
Aluminium–copper alloys
Mild steel, cast and wrought iron
Lead–tin solders
Lead
Tin
Brass
Copper
Bronze
Nickel
Stainless steels

Types of corrosion

Corrosion may be defined as 'the deterioration of a metal by chemical or electro-chemical reaction with its environment'. This deterioration takes place in different ways depending upon the corrosive medium, presence of other metals, and temperature, etc.

There are various types of corrosion. Some of the commonest are:

Pitting This is a localized random form. With aluminium its rate of penetration decreases with time, varying from alloy to alloy.

Galvanic This type of corrosion takes place when dissimilar metals are effectively coupled together in the presence of moisture or some other electrolyte. The severity of corrosion is dependent upon the relative position of the two metals in the 'galvanic series'. See Table 6.1. A metal is corroded galvanically by any metal beneath it in the table.

Crevice If aluminium is positioned such that a crevice is created between it and another metal, then crevice corrosion can occur. At the mouth of the crevice oxygen is freely available, whereas at the tip it is relatively sparse. In consequence electrolytic action can occur due to the different potential between the two regions.

Deposition This is a special form of galvanic corrosion in which ions of a more noble metal are deposited from a solution onto aluminium thus setting up electrolytic action resulting in pitting corrosion. Copper ions in aqueous solution are troublesome to aluminium, and for this reason it is always recommended that copper and aluminium should not be used together in building applications where, of course, damp conditions cannot be avoided.

Alloy choice

The term 'aluminium' is usually used as a reference to the commercially-pure metal (99.0–99.5 per cent purity).

By far the best corrosion resistance and durability are obtained where super-purity aluminium (99.99 per cent aluminium) is used, but this specification, because of its high extraction cost, is only rarely used in industry. Commercial-purity aluminium, and all of its many alloys, contain small amounts of other elements such as manganese, magnesium, copper, iron and silicon. It is these elements that determine the corrosion behaviour of the alloy, and as a general guide it can be stated that the more copper and iron contained the higher the susceptibility, and the smaller the amount of these two metals the lower the susceptibility.

The commonly used building sheet alloy (BS3103) has very good corrosion resistance. Of the stronger alloys the aluminium–magnesium series are very resistant, and are particularly good for marine environments.

Alloys containing magnesium and silicon are also satisfactory but the very strong copper-bearing alloys have poor resistance and are frequently given a cladding of pure metal to overcome this problem.

The aluminium–magnesium–silicon extrusion alloys have good weathering and durability, as have the aluminium–silicon and aluminium–magnesium casting alloys (Table 6.2).

Table 6.2 Durability and resistance to atmospheric attack

Alloy designation	Main alloying elements	Aluminium %	Resistance factor
1080 A	–	99.80	E
1050 A	–	99.50	V
1200	–	90.00	V
2000 series	Copper	Balance	P
3000 series	Manganese	,,	V
5000 series	Magnesium	,,	F–V
6000 series	Magnesium and silicon	,,	G–V
7000 series	Zinc	,,	F–G
8000 series	Lithium	,,	F*

E = Excellent   V = Very good   G = Good   F = Fair   P = Poor   * = More experience needed

Compatibility with foods and chemicals

Aluminium is resistant to many, but not all, foods and chemicals. It is an inert, non-toxic, odour-free metal and is widely used in food and chemical processing plants and in packaging foodstuffs.

The corrosion behaviour of aluminium vessels and equipment in service is influenced by many factors, not only the chemical nature of the product in contact with the aluminium but also the temperature, concentration, degree of agitation and aeration, and so on used during the processing. The likelihood of corrosion or otherwise, is also affected by the design of the plant. Galvanic and crevice corrosion, in particular, can be reduced or eliminated by careful design, such that dissimilar metal contact is avoided and corners and crevices that are difficult to clean are eliminated.

7 Aluminium and the environment

As the third most abundant element in the earth's crust, after oxygen and silicon, aluminium can justifiably be called one of the world's least scarce resources. It is also one of the easiest to recycle; used aluminium may be reconverted into high-quality metal in a very cost-effective operation. This combination of advantages is one of the reasons why aluminium is becoming regarded as a 'green' material, friendly to the environment and the people in it.

The potential energy-saving characteristics of the metal in action are also important in many industries, particularly transport and packaging. It can be shown, according to figures published within the aluminium industry, that in many applications the energy savings through using aluminium products (to reduce weight and save fuel consumption, for example) are greater than the additional energy required initially to make an aluminium product, rather than (say) a steel one.

This positive energy equation works to aluminium's advantage across a wide range of industries and is reinforced by the energy savings obtained by recycling aluminium. The conversion of scrap aluminium back into re-usable high-grade metal requires only 5 per cent of the energy needed to make the same weight of virgin metal.

So recycling aluminium is assuming a greater importance than in less energy-conscious times and the high recycle value of aluminium has become an important 'plus' factor in material-selection decisions and is likely to assume even greater significance in the future.

Within the aluminium industry itself recycling plays an important part

Aluminium scrap, from production offcuts, all finds its way back into the melting pot to be turned back into re-usable high-grade metal

in efficient plant operations. Constant recycling takes place of all process scrap from rolling and extruding such that a recycle rate of virtually 100 per cent takes place.

The variable element in scrap recycling concerns the recovery of discarded products such as saucepans, cans, packaging, motor-car components and so on. Much of this scrap, however, is easy to collect and a high percentage of 'consumer' scrap of this kind finds its way back into the production cycle. Perhaps the aluminium product that currently has the highest public profile regarding scrap recycling is the aluminium drinks can. Over four billion aluminium drinks cans are used and then discarded annually in the UK alone. Such a high volume of cans represents a large and valuable tonneage of re-usable metal and the UK aluminium industry is heavily committed to developing efficient collection and recycling.

It is not only consumer products that have a recycle value, however. For example, thousands of tonnes of aluminium are used every year for the production of window frames and many thousands more tonnes go into other building products. All of these products have a long, perhaps very long, working life, but at the end of that time they still have a high recovery value and represent a valuable source of high-grade metal.

As yet there is little, if any, recognition of this long-term asset value, when material choice is first considered. Of course, priority factors such as performance, cost-effectiveness, strength, durability, weight, operating temperature, environment and safety all have to be satisfied. But in addition, with conservation of materials and resources now so important, and with many ecological considerations also involved, this extra benefit offered by aluminium could become as important in the building industry as it is already becoming in packaging, transportation and other market sectors.

Part II
A BUILDING MATERIAL

8 Joining the establishment

Aluminium has gained itself a place alongside other materials such as brick, concrete, steel and timber as a building material with proven characteristics and behaviour. The unique combination of properties, versatility of form, light weight and durability have made aluminium a building material that is ideal in modern construction. Today's designs of tall, strong but light-weight commercial buildings make important use of aluminium cladding and curtain walling with its material benefits coupled with the ease and speed of erection. Domestic houses incorporate double-glazed aluminium window frames that offer a lifetime of attractive and efficient service.

But curtain walling and windows are just two examples of aluminium in action. The varied characteristics of the metal have opened up uses of rolled metal from thin aluminium foil for vapour and moisture barriers in insulation panelling to thick treadplate for walkways and of extruded sections for products from ladders to space structures. The versatility of aluminium is well demonstrated by its vastly differing market uses, yet all in specific ways taking advantage of aluminium's properties of strength, lightness, durability and formability.

Aluminium's progress into, and acceptance by, the building industry was hindered in the early years of the metal's growth in the twentieth century by the dominance of the huge, longer-established steel industry, and aluminium was compared unfavourably – it was softer, more expensive and bent more easily. However, the experience gained using aluminium in the aircraft industry from 1940 onwards changed these attitudes.

Flexible piping made from crimped aluminium foil, sometimes in combination with kraft paper, is useful for a variety of heating and ventilating applications

Appreciation of the properties of the many alloys; realization that for a metal only one-third the weight of steel cost/tonne comparisons are meaningless; and importantly, recognition that aluminium is a metal having its own structural characteristics calling for a new design approach and not a steel-copy approach, have all resulted in aluminium now being treated as a building and engineering material in its own right with its own specific uses and advantages.

Early pioneering developments with aluminium in the 1930s, notably in aluminium windows, were followed by the major production at the end of World War II of the aluminium prefabricated bungalow as a means of providing rapidly-erected, temporary housing, using existing assembly-line facilities no longer required for aircraft manufacture. This was the first time in this country that mass-production techniques had been applied to housing production. The walls of the 'prefab' as the bungalow became known were made of aluminium sheeting with an infill of trapped-air cement grout. Structural members too were made of aluminium, and a total of about 1¾ tonnes of sheet, strip, castings and extrusions was used in each prefab. Made from aircraft scrap material, the alloy used contained a high copper content which would not nowadays be recommended for building applications because of this alloy's susceptibility to corrosion. But, despite some problems caused by insufficient protective painting, the bungalows were remarkably successful. In fact a large number of these prefabs were still in use many years after their envisaged timespan of 10 years' occupation, and at the time of writing some are still in use 45 years later.

This revolutionary and successful use of aluminium in housing marked the start of an extensive, and intensive, research and development pro-

Strong, light and durable, aluminium alloy ladders are now commonplace and have widely displaced wood

gramme by aluminium companies in the UK to develop previously unheard-of uses for aluminium.

Today, as a result, aluminium is well-established in a vast number of product areas ranging from major structural applications such as space frames and patent glazing to roofing and cladding, shop fronts, curtain walling, windows and conservatories. Other important applications include gutters and downpipes, flashings and copings, ducting, handrails, bridge balustrading, road signs, venetian blinds, insulation panels, door furniture, grilles, partitioning and ceiling panels.

Some castings and forgings are used for a wide variety of building components, but the bulk of metal used in building is in the extruded and rolled forms. Over 70 000 tonnes of extrusions and 25 000 tonnes of sheet go annually into this important aluminium market in the UK. The size of

It is not just for standard non-structural window frames where aluminium finds many uses. Here is a security window undergoing severe ballistic testing

the market for extrusions is evidence of the versatility of the extrusion production process for obtaining shapes that are custom-designed to do a specific job and to do it in a highly cost-effective manner. Not only can shapes be made that include many features in one integral section, but also the metal can be placed where it is needed and omitted where it is not needed. This facility is very cost-effective, enabling metal weight to be minimized and many subsequent machining, drilling and routing operations that might otherwise be necessary to be eliminated.

The range of alloys available gives the designer of aluminium components a comprehensive choice. A small selection from the many available provides a means of covering the whole range of likely building applications. For example, commercial-purity metal (99.0 per cent minimum

A unique appearance was given to the Sainsbury Centre for Visual Arts at the University of East Anglia by using 2 500 cladding panels shaped using super-plastic alloy sheet (courtesy SuperForm)

aluminium), known as alloy 1200, is in sheet form a perfectly suitable choice for flashings, louvres and other components where very easy formability is important but strength is not. For applications such as roofing and cladding where some additional strength is needed and where good durability and resistance to atmospheric attack is required then alloy 3103 containing about 1 ¼ per cent manganese is considered best.

For extruded sections requiring easy extrudability in complex shapes, reasonable strength, a good surface finish and high resistance to atmospheric corrosion, alloy 6063 containing magnesium and silicon is universally acknowledged as ideal. Applications requiring to be more highly stressed are generally suited by extrusion alloy 6082, which is similar to 6063 but gains added strength by a slightly higher magnesium and silicon content and the addition of a little manganese. The very high-strength alloys containing copper and zinc are not generally recommended for structural building applications because of their lower corrosion resistance. These high-strength structural alloys are, however, successfully used for military applications such as bridges and pontoons, where the combination of strength and light weight enables structures to be readily erected and dismantled.

Interest is also being shown in Al–Zn–Mg alloys for offshore building

A continental European example of extruded aluminium sun louvres

applications where the use of light-weight, above-sea level structures such as accommodation units is becoming increasingly important.

Throughout the building industry, therefore, aluminium has over the 100 years of its commercial life moved from a novelty metal to a material that has taken its place alongside traditional building materials such as stone, brick, wood, glass and steel. The early use of aluminium was, as might be expected, for various decorative and statuary applications. At that time no reliance could be placed on the metal's strength and durability; confidence in these attributes has had to develop over years of use. One of the first recorded uses of aluminium in building is the pyramid capping on the Washington monument in Washington DC, which was completed in 1884. In the UK the handsome statue of Eros in Piccadilly Circus assembled using separate castings in 1893 is an early and still standing example of aluminium statuary.

Inside and outside modern buildings aluminium is now playing its part in shaping and styling modern architecture, providing both grace and functionality to the many tall buildings that dominate our city skylines and to the local high street where aluminium shopfronts abound in colourful juxtaposition.

Military bridges are made from high strength Al–Zn–Mg alloys. Armoured vehicles make good use of the same material

9 General design data

Forms and sizes

Aluminium in the form of castings, forgings, plate, sheet, strip, foil, extruded sections and tube are all manufactured in the UK and are available direct from the manufacturers. Many of the forms are held in stock by the wide network of UK stockists. Stockists not only offer just-in-time delivery service, but also provide comprehensive cutting, forming and finishing services.

Castings

Depending upon the production method used castings can be produced weighing just a few grammes or 1 000 kg or more.

Plate

The maximum width to which plate can be produced is 3 500 mm*. Plate length and maximum thickness is governed by the starting size of the rolling slab, which can be up to 12 tonnes. Stockists hold a variety of plate sizes and thicknesses, and some specialists offer cutting and shaping services. Typically the maximum size of plate available from stock is 12 ft × 6 ft (4 m × 2 m).

Sheet

The maximum practical sheet width is 2 m.* Standard sizes up to around 12 ft × 6 ft (4 m × 2 m) are generally available through stockists in a range of thicknesses. In practice, for both sheet and plate materials, handling considerations dictate the maximum practical size available.

Strip and coil

The maximum width applies as for sheet.

Patterned sheet and treadplate

Sheet is available in a range of thicknesses and sizes with impressed decorative patterns. Stucco-embossing which gives a 'mottled' effect is commonly used on building sheet to reduce glare and light reflectivity. Other patterns are available to provide a variety of decorative effects.

Treadplate made from heavier thickness sheet or plate has sharply raised patterning that provides a good non-slip surface.

Extrusions

Most extruders offer widths up to around 9 inches (228 mm), but shapes are available using large extrusion presses up to a maximum width of about 30 inches (800 mm).

*(*Note* For specific information on available sizes, readers should refer to the manufacturers. The details given here are intended as a guide only. Not all manufacturers will make to the maximum sizes indicated.)

Behaviour in service

Aluminium owes its excellent weathering characteristics to the tough, adherent oxide film that is always present on its surface when in contact with, or following any contact with, the atmosphere. This film, which originally is less than 0.01 μm thick, re-forms spontaneously in air should the surface be cut or abraded.

The atmospheric protection offered by this oxide film varies with its thickness and the chemical composition of the alloy. Pure aluminium and the medium-strength alloys are more resistant to atmospheric attack than the high-strength alloys containing copper and zinc. All alloys, however, where the surfaces are kept clean and dry remain virtually unchanged,

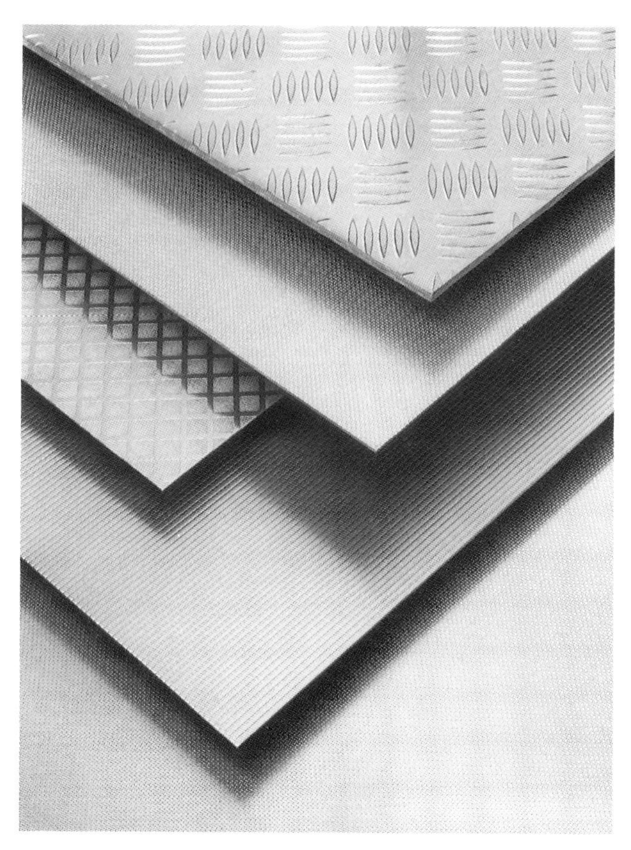

Rugged 5-bar treadplate and various patterns of decoratively embossed aluminium sheet

but in damp conditions the oxide film starts to thicken up and in so doing develops a roughened dull appearance. This increase in thickness provides additional protection to the underlying metal reaching a maximum after about two years' exposure.

In aggressive environments, such as marine and particularly polluted industrial areas, the rate of oxide formation may be increased with a consequent adverse effect on appearance. Importantly, this increase in oxide thickness, which tends to be self-stifling in time rather than progressive, has no significant effect on the metal's structural performance. The enormous experience built up over the past 60 years and more has demonstrated the excellent durability of the commonly used building alloys.

Today's practice is to anodize or paint much of the aluminium sheet and extrusions used in building and this not only improves the decorative appeal by adding colour, but also adds many years' service to the surface durability and appearance.

In all cases the best performance from aluminium in building is obtained by:

1 Regular cleaning wherever practical and possible.
2 Allowing rainwater to wash exposed surfaces.
3 Avoiding static moisture and condensation.
4 Avoiding contact between aluminium and other metals, particularly copper.
5 Designing-out dirt and moisture traps.
6 Designing-in moisture escape routes.

Contact with water

Rainwater

Direct washing of aluminium by rainwater is beneficial, but water should not discharge onto aluminium after passing over a copper roof or through copper pipes.

Domestic water

Domestic waters vary greatly in their salt and mineral contents. Under certain conditions these can cause pitting corrosion of aluminium, and so aluminium piping is not generally recommended. Closed-circuit systems in which the aluminium is treated or which include an inhibitor in the water are generally acceptable.

Seawater

Aluminium building alloys are resistant to seawater corrosion and aluminium is accepted as a good coastal region building material.

Other waters

Aluminium is a popular choice for covering swimming pools and small reservoirs. The strength and lightness of the metal are advantageous for covering wide spans with a minimum of supporting structure. The small chlorine content in the water of swimming pools has no harmful effect on an aluminium roof. Ventilation, however, is essential to remove water vapour and minimize condensation.

Contact with building materials

Cement and mortar

Dry cement and mortar in contact with aluminium have no harmful effect. However, contact with wet cement, concrete or mortar will attack the surface of contact aluminium, but the rate of attack drops rapidly as drying out occurs. As a precaution therefore against corrosive attack in damp conditions any surfaces of the metal in contact with the cement and mortar should be protected with a coat of bitumen compound.

It should also be noted that anodized surfaces can be stained by droppings of wet cement. A protective lacquer or peelable paper covering the metalwork during construction operations is therefore useful.

Timber

Aluminium and timber can be safely used in contact. It should be noted though that some timber preservatives may be harmful to the metal, and where preserved wood is in close proximity to aluminium particularly under conditions of high humidity, it is advisable to coat the touching aluminium surfaces with a protective paint or bitumen.

Plastics

Aluminium and plastics can be safely used in contact.

Metals

Electrolytic action takes place when two dissimilar metals are in contact under damp conditions. The metal which is the more electro-negative of the two is corroded while the one that is more electro-positive is protected. The galvanic series (see Table 6.1, p. 64) lists the order of nobility of the metals with the least noble (most negative potential) magnesium heading the list. At the bottom of the list come stainless steels suggesting that these steel alloys are to be avoided with aluminium. This in practice is *not the case*. The surface oxide film on stainless steel causes a change in polarity which makes it safe to use stainless steel fixings with aluminium except in marine and aggressive industrial environments.

The safest fittings to use are those that are zinc coated or galvanized.

Mild steel fittings should preferably be given a barrier coating of a suitable paint.

Finishes on aluminium

Anodizing

Anodizing is used to provide both additional protection to aluminium and to give a choice of unique colourful finishes. Various anodizing processes specially developed to suit architectural needs are available. These provide the specifier with a range of colour options through from natural silver to greens, blues and reds, to shades of bronze and to black. The thickness of the film determines its durability and BS1615 and BS3987 lay down specifications.

Painting

Prepainted sheet, strip, coil and extrusions are widely used to provide the specifier with an option of colours as broad as the range of paints permits. Prepainted aluminium may be formed, cut, drilled and fabricated in the same way as mill-finish metal with no harm being caused to the coating provided correct care is taken. Alternatively, many components are coated after forming under controlled factory conditions. Finally, on-site painting is always a possibility.

Handling and storage

Aluminium products need to be handled carefully to avoid surface damage. Ideally, aluminium components should be dry on arrival on site. It is possible, however, that they carry surface condensation, and in the case of a consignment of sheet, the condensate may have penetrated between the individual sheets by capillary action. This might cause surface staining and it is therefore advisable to check the material on arrival.

Storage conditions should be warm, dry and free from condensation. Where possible extruded sections should be stocked vertically rather than horizontally and sheet should be stacked on edge rather than flat. Battens should always be used to avoid damage to the aluminium by contact with rough ground.

Building materials such as cement, plaster and cleaning acids should be prevented from splashing onto aluminium surfaces.

Cleaning and maintenance

Aluminium is a highly durable metal and it has been common practice to regard aluminium as a 'no-maintenance' building material. It is certainly the case that, left untouched for years, an aluminium component under most atmospheric conditions will remain structurally sound and will not deteriorate in a manner likely to affect the performance of the component. It is not therefore necessary to refinish aluminium surfaces on a regular basis because of any fear that the metal will corrode away if not protected, but it is necessary to apply cleaning maintenance if the surfaces are to retain a clean attractive appearance. Rainwater will play its part in cleansing exposed surfaces but regular cleaning with water and a little detergent, followed by rinsing with clear water, is recommended to remove atmospheric dirt and pollutants.

Waxing the surfaces, after cleaning and drying, is an additional protection, and one that is beneficial for windows and shop-fronts.

Neglected surfaces cannot be restored to their original condition. Badly neglected mill-finish metal will roughen and darken and perhaps exhibit localized pitting corrosion. Such surfaces can be cleaned with a mild abrasive such as a metal polish or impregnated nylon pad. Coarse abrasives and metallic wire-wool scouring pads should never be used.

Whether an anodized finish, natural or coloured, or an organic paint coating is chosen, many years of excellent performance may be expected provided that simple cleaning on a regular, albeit infrequent, basis is carried out. With many applications this is most easily done when the window panes are cleaned. No one expects windows not to need cleaning and the same is true of their surrounds.

Part III
APPLICATIONS IN BUILDING

10 Windows and doors

Windows

The first aluminium windows in the UK were installed in the early 1930s. Natural anodized frames, basically copies of existing steel casement windows, for example, were fitted to the New Bodleian Library, Oxford, and in a new office block in Banbury. Both of these installations are still in good condition today.

Since then designs have become much more sophisticated and specific to the aluminium extrusion process. Today's aluminium frames, smooth and sleek on the outside visible surfaces, have inbuilt internal features that provide a window performance undreamt of in the 1930s. Retaining grooves for weatherstripping, screw ports, condensation channels, interlocking devices, grooves for butyl strip and thermal insulation channels are all as-extruded features of the various sections that together make up a suite of window sections.

The early casement style has been superseded by a range of styles to suit differing requirements (Figure 10.1). Fixed lights and opening lights that include vertical and horizontal sliders, horizontal and vertical pivot, top hung, bottom hung, side hung, and tilt 'n' turn designs offer today's designer a choice for all buildings. Complementary subframes, too, are available in aluminium, plastics and hardwood, although these are not essential as many aluminium frames can be fixed directly into brickwork and masonry (Figure 10.2(a) and (b)).

The usual alloy selected for window frames is BS1474/6063. Both solid

The offices of British Alcan Extrusions at Banbury were fitted with silver-anodized aluminium windows in 1935. Fifty-five years later they are still in excellent condition (photo taken 1990)

and hollow sections are used, and many of the solid sections are designed to clip together forming a frame that is multi-hollow with easy access to interior components.

The strength of alloy 6063 is such that slim frames are the norm for aluminium windows, maximizing the amount of glass that may be used.

Not only do aluminium frames, either painted or anodized, possess excellent durability, they also are immune to warping, twisting, sticking and rotting. Aluminium alloy sections retain dimensional accuracy and stability.

The market for aluminium windows, certainly in the domestic area, has been coincident with that for double-glazing. Aluminium frames accept double-glazed panels making a combination that provides high-performance windows.

Thermal insulation of aluminium frames is very popular. Aluminium, a good heat conductor, will, if not insulated between inside and outside surfaces, provide a path for heat to travel from one side of the frame to the other. The phenomenon can manifest itself under appropriate conditions by condensation on the interior surface of the window frame. By inserting an insulating barrier either on the outside of the frame or in the inside this condensation can be minimized. The thermal barrier also prevents heat loss from the building to the outside, an important consideration in an energy-conscious age. It must be said, however, that the amount of heat

Aluminium was used extensively in the fenestration of the Yvonne Arnaud Theatre, Guildford

that can escape through an aluminium frame, as compared for example with the loss through a single-glazed instead of a double-glazed panel, is very small.

Methods of thermally insulating aluminium frames fall basically into three categories:

(a) Thermally clad The application of a plastic covering to the outside of the frame (Figure 10.2(c)).

(b) Thermally-broken The insertion of an insulating barrier into the centre of a section (Figure 10.2(d)). The general method of doing this is the Azon 'fill and de-bridge' method or similar. An alternative to this method, which takes one section and effectively separates it into two, is to take two sections and join them with a solid strip or strips of insulating plastic such as polyamide. The use of solid strip insulant provides the

Figure 10.1 Window styles available in aluminium

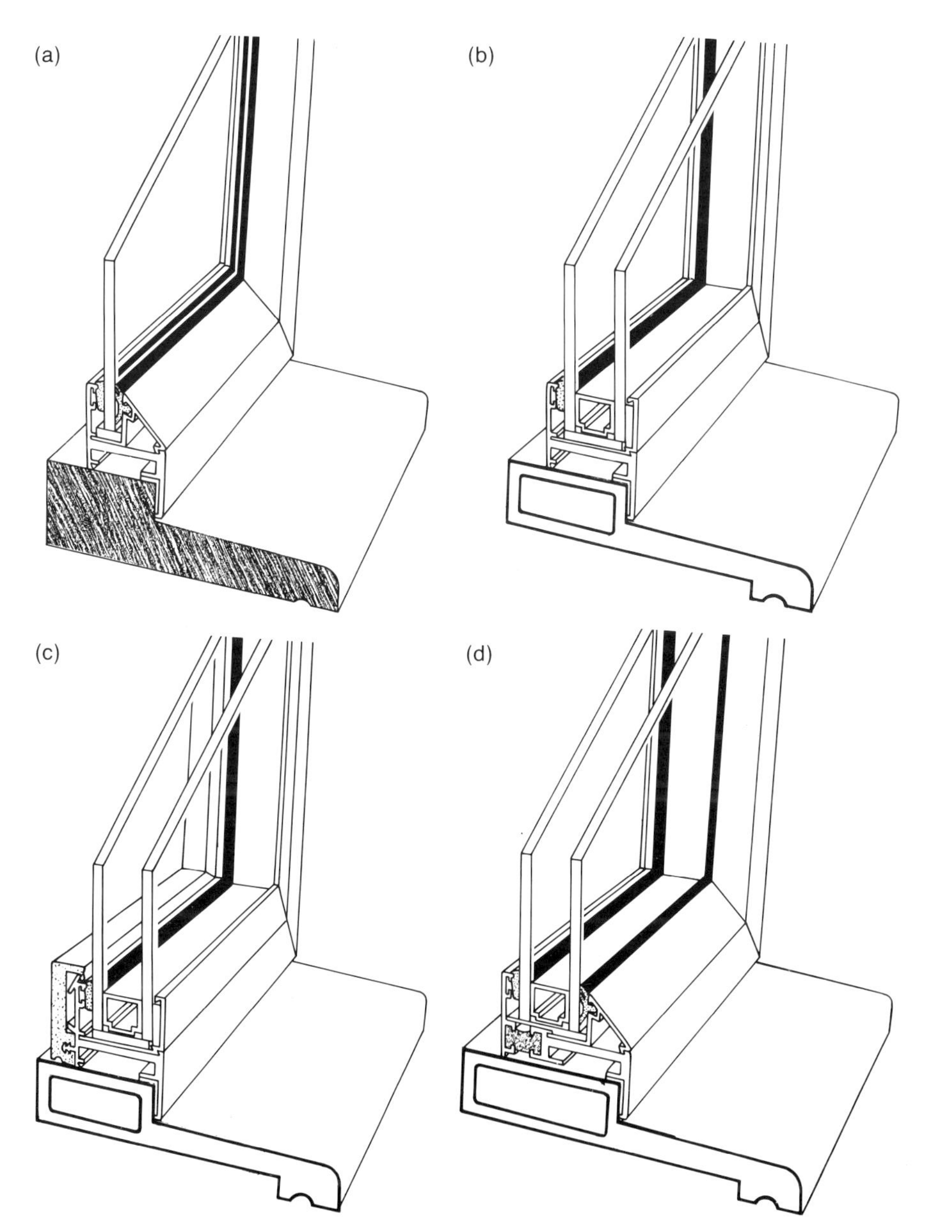

Figure 10.2 Window frames: (a) a typical aluminium single-glazed window frame set on a timber subframe; (b) a typical double-glazed, non thermally broken aluminium window frame set in a plastic subframe; (c) a typical thermally clad, double-glazed aluminium window frame; (d) a typical thermally broken, double-glazed aluminium window frame

This modern house is fitted throughout with white painted, double glazed and thermally broken aluminium windows and doors. Diamond-patterned leaded glass provides a traditional look (courtesy Kaye Aluminium)

additional benefit of allowing the use of two-colour effects. By joining two aluminium profiles, each of a different colour finish, it is practical to have a different colour on the inside and outside of the frame.

(c) Composite frames An increasingly popular trend today is to combine aluminium with either plastics or timber. This is a means some manufacturers claim of 'getting the best of two materials'. The combination certainly provides good insulation and additionally offers the strength and durability of aluminium as an external material exposed to the elements with the option of alternative materials and colours on the inside.

The use of aluminium as a framing material enables high-performance specifications to be achieved. Weatherstripping for example, such as woven pile, neoprene, pvc or ethylene propylene can be housed within retaining grooves extruded into the flanges of the sections. Pressure-equalization and self-draining features can also be designed-in to the frames. Figure 10.3(a), (b) and (c) illustrate composite frames.

Rigid and reliable corner construction is achieved by mitring and mechanically cleating. Although welding of aluminium is commonly practised it is only rarely used for the corner jointing of window frames.

Not only rectangular but curved frames are possible using aluminium

The slim aluminium frames in the thermally broken, double-glazed windows installed by South Herefordshire District Council provide a light, airy look to the cottage-style property. Minimum maintenance and long life were reasons for the installation (courtesy Warmshield)

Close-up of a pivoted thermally broken aluminium window equipped with a multi-locking device (courtesy Heywood Williams)

sections bent on modern equipment. Curves may be either in the plane of the glass or at an angle to it.

Doors

The strength and lightness of aluminium is particularly valuable for entrance doors of all kinds due to the demanding service conditions and traffic requirements to which doors are subjected.

Typically doors are constructed of hollow extruded sections in alloy 6063T6. The use of hollow sections not only provides a combination of lightness, strength and rigidity, but importantly, also enables many features such as multi-locking devices, bolts, spring closers and various security features to be concealed within the frame.

Options such as midrails and kick plates are easily incorporated and thermal-break features too are available.

Installing an aluminium 'picture' window

Georgian-style aluminium white painted windows (courtesy Caradon Everest)

These black powder-painted thermally broken windows with leaded double glazing give a distinctive look to this cottage style property (courtesy Caradon Everest)

This 3-storey extension to Christ College, Brecon, is fitted with white-painted thermally broken aluminium windows. Note the top centre curved frame above the entrance door (courtesy Smart Systems)

Patio doors

The patio door market is worthy of individual mention on two counts. First it was, historically, the starting point for aluminium's rise to promi-

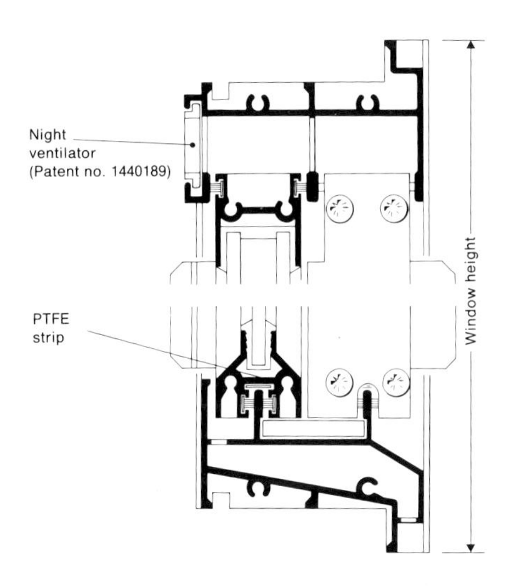

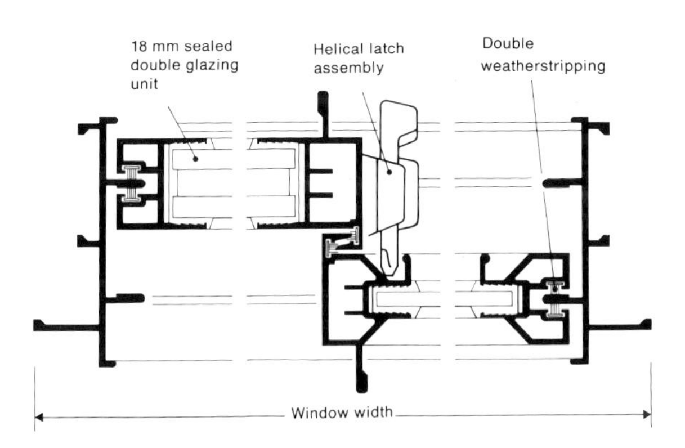

Figure 10.3 Composite frames: (a) details of a horizontally sliding window (non thermally broken) (courtesy Avdon)

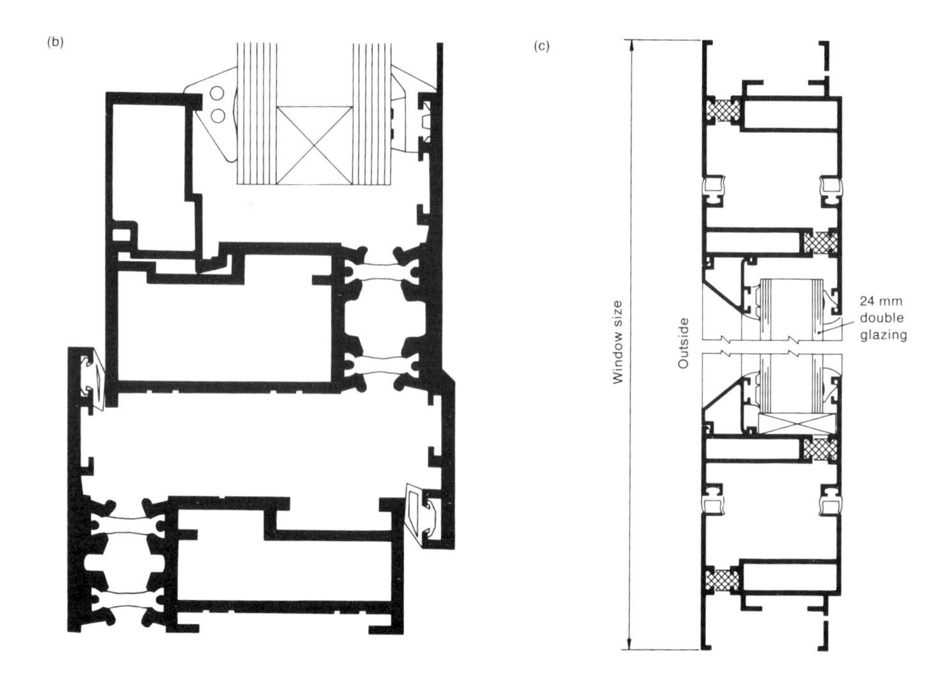

Figure 10.3 (Continued): (b) typical detail of a casement window incorporating polyamide thermal break strips. These permit the inside and outside sections to be finished in different colours (courtesy Kawneer Europe); (c) typical cross-section of a thermally broken pivoted window (courtesy Heywood Williams Windows, Chester)

nence in the domestic home improvement market, and second it remains one sector where aluminium has been least affected by the emergence of the plastics material PVC–U (based on polyvinyl chloride) as an alternative, competitive material.

It is also the one area where aluminium is a major choice in 'new-build' housing as well as in the home improvement market. Whereas currently soft wood overwhelmingly dominates the new-build window sector, with aluminium taking only a small share, the reverse is true of sliding patio doors where aluminium is the first choice, with wood only taking a small part of the market.

High standards of safety, security, weather performance, insulation and slim appearance are all achievable with aluminium frames, which include such items as specially designed roller tracks, inbuilt double weather strip retention grooves, and pressure equalization features.

This white-painted aluminium door frame is fitted with a contrasting gold anodized lock handle and letter flap

Contrasting black and silver anodized finishes help provide an attractive entrance to this commercial building

Aluminium patio doors remain one of the most popular uses for aluminium in domestic fenestration, either for new or improved properties

Many different types of fittings and fastenings can be designed into the frames, such as hook latches, multi-point fastenings and espagnolette bolts. Figure 10.4(a) to (d) illustrate sections of various doors and patio doors.

Security

The rapidly rising necessity to protect property from unwanted intrusion has led to many sophisticated locking devices being incorporated into aluminium window and door designs.

Additionally there is a growing trend towards the use of exterior aluminium shutters and security screens. Wooden shutters, of course, are commonplace in Europe, where windows are designed to open inwards rather than outwards as in the UK. Now it would appear that the concept of shuttering on windows is gaining momentum in the UK, based not on wood, but on a combination of aluminium sheet and extrusions. There is significant future potential for aluminium in this new market area.

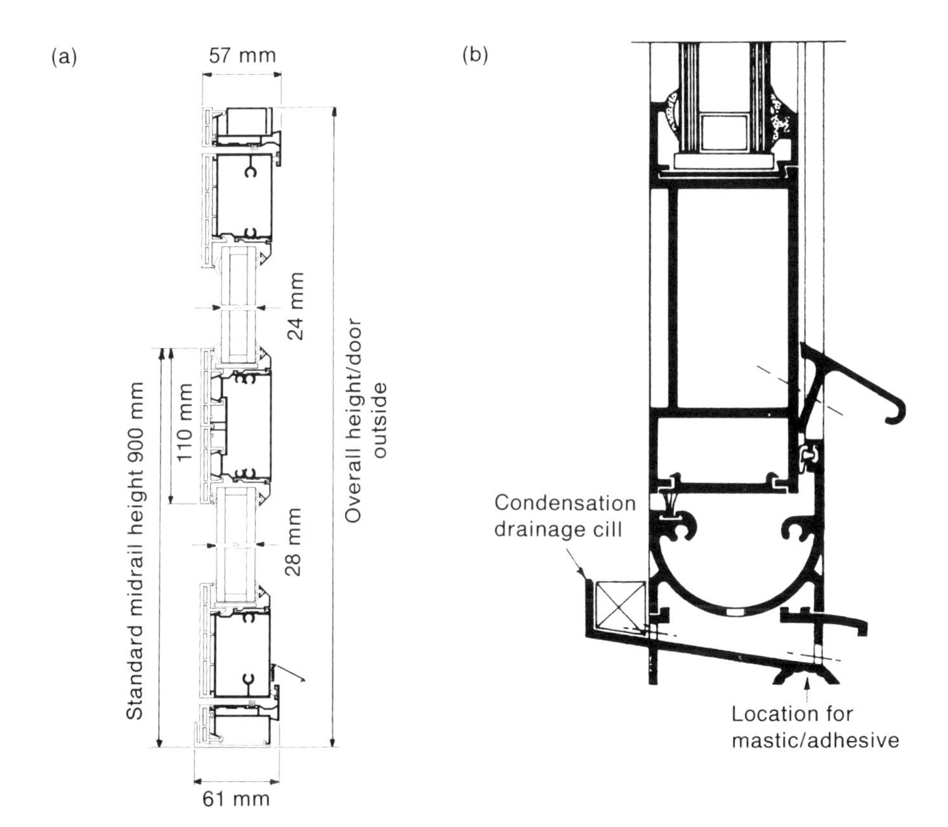

Figure 10.4 **Various doors and patio doors: (a) detail through a composite plastic-aluminium residential door frame (courtesy Smart Systems Duotherm); (b) typical cill detail of non thermally broken, double-glazed residential door (courtesy Tru Architectural Products)**

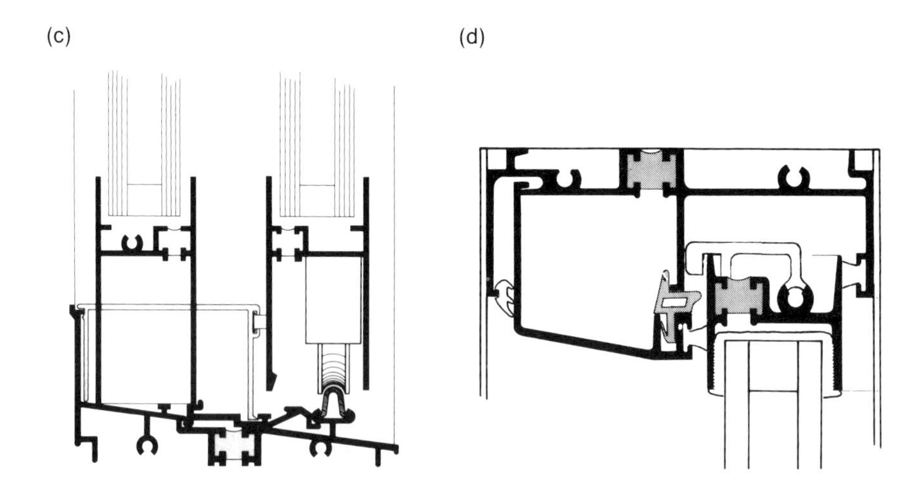

Figure 10.4 **(Continued): (c) typical detail of sliding patio door (thermally broken) (courtesy Kaye Aluminium); (d) section through sliding panel of thermal barrier patio door (courtesy Glostal)**

11 Curtain walling

Historical development

The need to clad a building structure has existed since the earliest recorded times. The aboriginal tents of North America and South Africa, for example, illustrate how cladding to withstand elemental conditions was achieved with primitive materials and tools.

In more recent times, the nineteenth century was dominated by the evolution of ironwork as the structure and glass as the cladding material. Based on these two materials 'frame and fill' techniques, as they were known, became successful architectural practices for many decades.

What is now commonly called curtain walling is a much more recent development which follows on from the pioneering work in the USA around the 1880s and 1890s when the skyscraper architects developed structure and cladding to a degree unknown elsewhere in the world.

The arrival on the commercial scene in 1888 of aluminium led logically to the use of this lightweight metal for the cladding of buildings, which, coupled with the increasing availability of glass panes in larger sizes and bigger production quantities spurred the development of this new approach to cladding. With load bearing concentrated in the structural framework of a building, the size of openings was no longer restricted by wall strength and so the design approach changed dramatically.

In theory thin, lightweight, easily-hung aluminium wall cladding enables structural members to be reduced and floor-space to be increased. In practice, the maximum theoretical savings are generally not achieved,

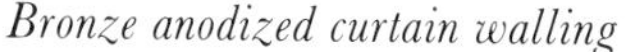

Bronze anodized curtain walling

The Euston Central building, finished in 1969, emphasises its high towers with naturally anodized mullions

but real advantages are certainly obtained. Rapid erection, accuracy and precision of component sizes, guaranteed performance, quicker internal finishing and also earlier building occupancy all follow from the ability to 'skin' a building with pre-made factory units.

Types of curtain wall

Curtain walling, which may be defined as a light cladding that covers and protects the sides of a building, requires to be fixed to the building structure. (Table 11.1 describes the functions of curtain walling.) This secondary means of support can take various forms such as a lightweight framework (generally aluminium) erected between or over the primary structural frame; a bracing frame contained within the cladding itself; or a fixing device attached directly to the structure. In all cases the use of aluminium extrusions assists in the provision of a lightweight but strong framework.

The cladding itself may consist of both a framing system of mullions and transoms and panel components either of glass or opaque sheet material such as aluminium.

Table 11.1 Function of a curtain wall

Function of a curtain wall
To withstand:
The action of the elements
Wind
Rain, snow, hail, sleet
Sunlight
Temperature variations
Atmospheric pollution
Fungal, insect and vermin attack
To prevent:
Unwanted access
Ingress of moisture
Fire damage
Injury to occupants
To control:
Heat transmission
Air movement
Light
Sound

The advantages of fast track curtain wall construction, which eliminates the need for scaffolding, enabled City House, Isle of Dogs, London, to be erected economically and rapidly (1989)

Close up of aluminium curtain walling

A prefabricated aluminium curtain-wall panel being lifted into position during the construction of the Forum Hotel, Glasgow, 1989

The North London Blood Transfusion Centre (1989) contains offices, laboratories, storage and garaging. Externally the building is clad in a mixture of glazed curtain walling, continuous horizontal windows and aluminium faced cladding panels. Inside, aluminium is used very extensively for shelf systems, fire resistant screens, partitioning, handrails, internal glazing, ceiling panels, ducting and laboratory furniture

The curtain walling may be either built up on site in what is called 'stick' construction or built up in pre-assembled units under controlled factory conditions ready for easy and rapid installation on site.

The use of aluminium extrusions for mullions, transoms and other components offers all the same advantages to a curtain wall construction as to windows and doors. As with these latter products, many curtain wall systems incorporate thermal-break features in order to minimize heat loss and condensation.

Aluminium faced infill panels are usually fitted with an insulation liner, sometimes fitted on site, but mostly forming part of the cladding specification offer to the curtain wall manufacturer.

The use of aluminium-foil honeycomb enables extremely lightweight but very strong robust panels to be constructed. Such panels can be faced with a variety of materials including not only flat, curved or shaped aluminium sheet, finished to suit the specific requirement, but also more unusual materials such as granite and marble facing.

Figures 11.1 and 11.2 show detail of two curtain wall designs.

The Merry Hill Centre, near Dudley, Worcestershire, makes extensive use of aluminium. The marble-faced curtain walling on this Debenham's store is backed with aluminium foil honeycomb to provide a strong, rigid and lightweight panelling (1990)

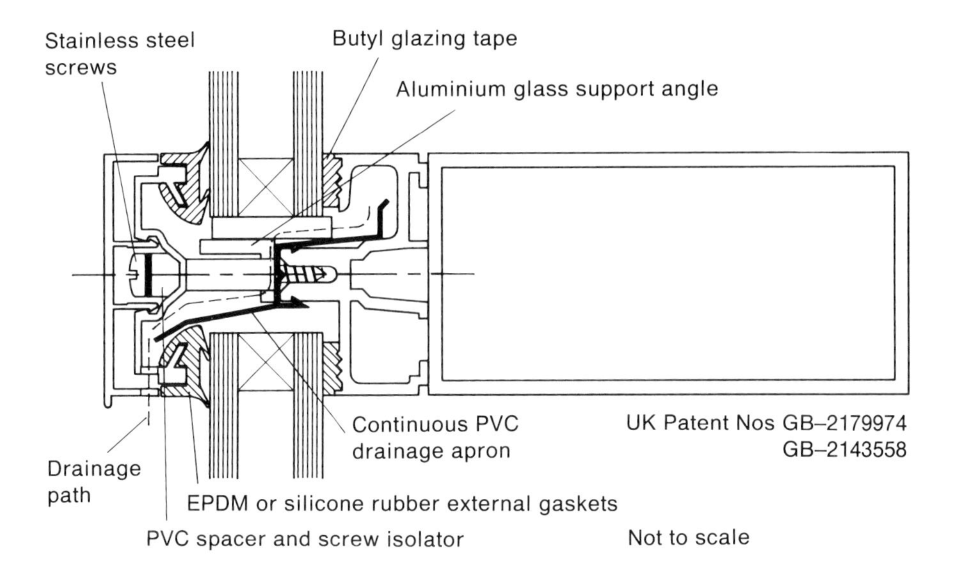

Figure 11.1 Curtain-wall design: detail of a pressure-equalized design with patented internal drainage arrangement (courtesy Pearce & Cutler)

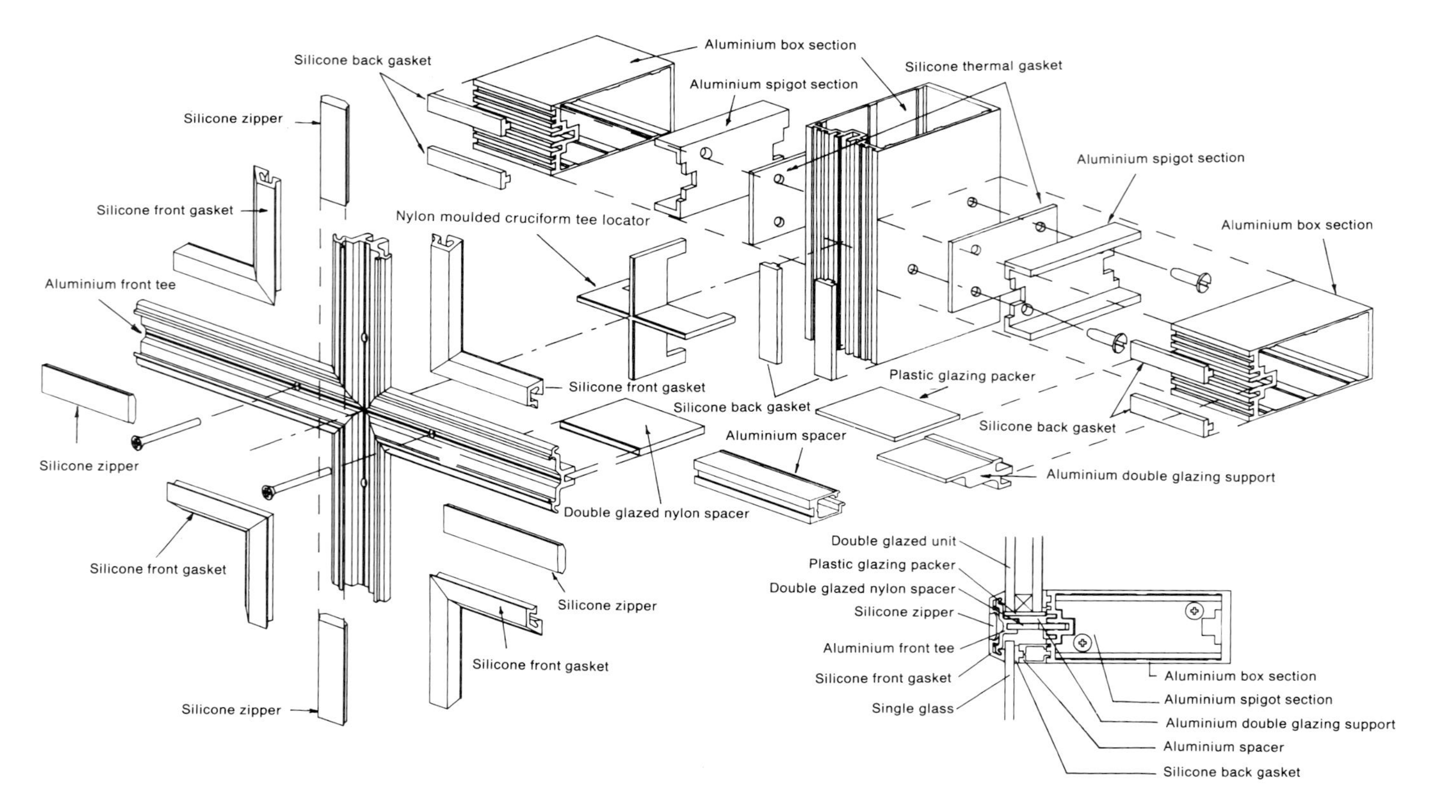

Figure 11.2 **Curtain-wall design: isometric view of a patented system using a silicone rubber gasket compressed by screwed aluminium glazing beads to secure the glazing and cladding panels (courtesy Stoakes Systems)**

12 Cladding and roofing

Wide range of uses

Aluminium profiled sheeting has been used for the roofing of industrial, commercial and agricultural buildings for many years. The earliest examples, at a time when aluminium was a relatively new material trying to break into markets held by other materials, were copies of corrugated iron designs. The sinusoidal pattern so popular with corrugated iron reigned supreme. Since then aluminium roofing has moved much more 'up market' in many senses – its profiled shapes are more varied and purpose-designed to accommodate the strength and rigidity characteristics of aluminium; it is provided in many designs including special secret-fix varieties and it is available in a range of thicknesses and finishes. The use of prepainted material is a current-day standard, with mill-finish metal generally used only on non-visible areas. Applications too, have broadened dramatically in scope, ranging from power stations to leisure centres, airport terminals, office buildings and mosques.

The need to provide claddings that not only are weatherproof but that also offer insulation against heat and sound transfer has become important. To meet this demand many composite panels are now available that incorporate insulant layers such as polyurethane foam or polyiso-cyanurate. To complete the sandwich, linings of aluminium foil or sheet are frequently included. Figures 12.1, 12.2 and 12.3 show details of aluminium sheet, secret-fix aluminium roofing and typical composite panels respectively.

Most profiles are produced by roll-forming, but in some cases brake-press equipment is employed. Forming techniques exist to produce curved profiled

Aluminium pre-painted cladding played an important role in the construction of the new Daily Telegraph building in London's Docklands (1988)

The International Garden Festival, Liverpool, was clad with profiled aluminium sheeting

The Bintulu Mosque, Malaysia, incorporates 'midnight blue' pre-painted aluminium sheeting supplied as coil from Alcan Duralcote (UK) to the Malaysian roofing contractors

Side-wall profiled cladding on a sports hall in Lancashire

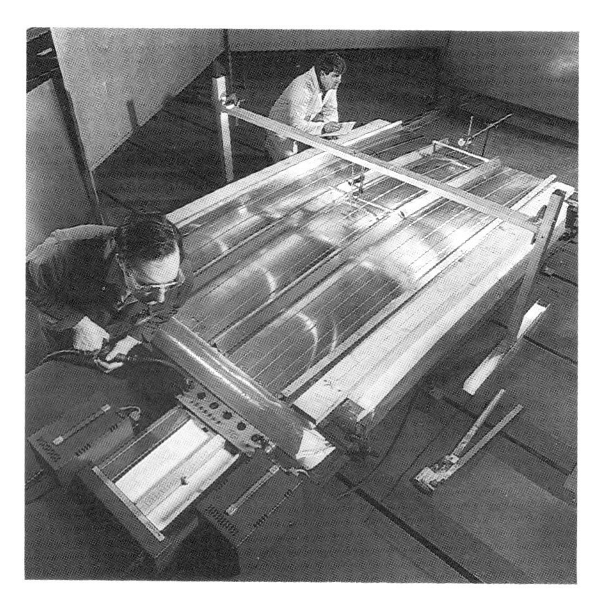

A length of profiled sheeting undergoing performance testing

Close-up of foam insulated aluminium roofing panels

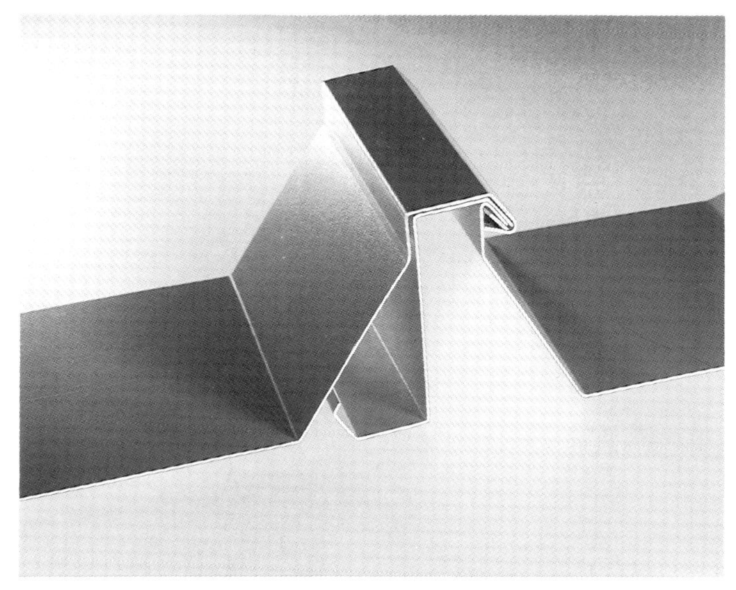

Detail of a hidden-fix roofing sheet system (courtesy British Alcan Building Products)

Aluminium roofing being fixed in position

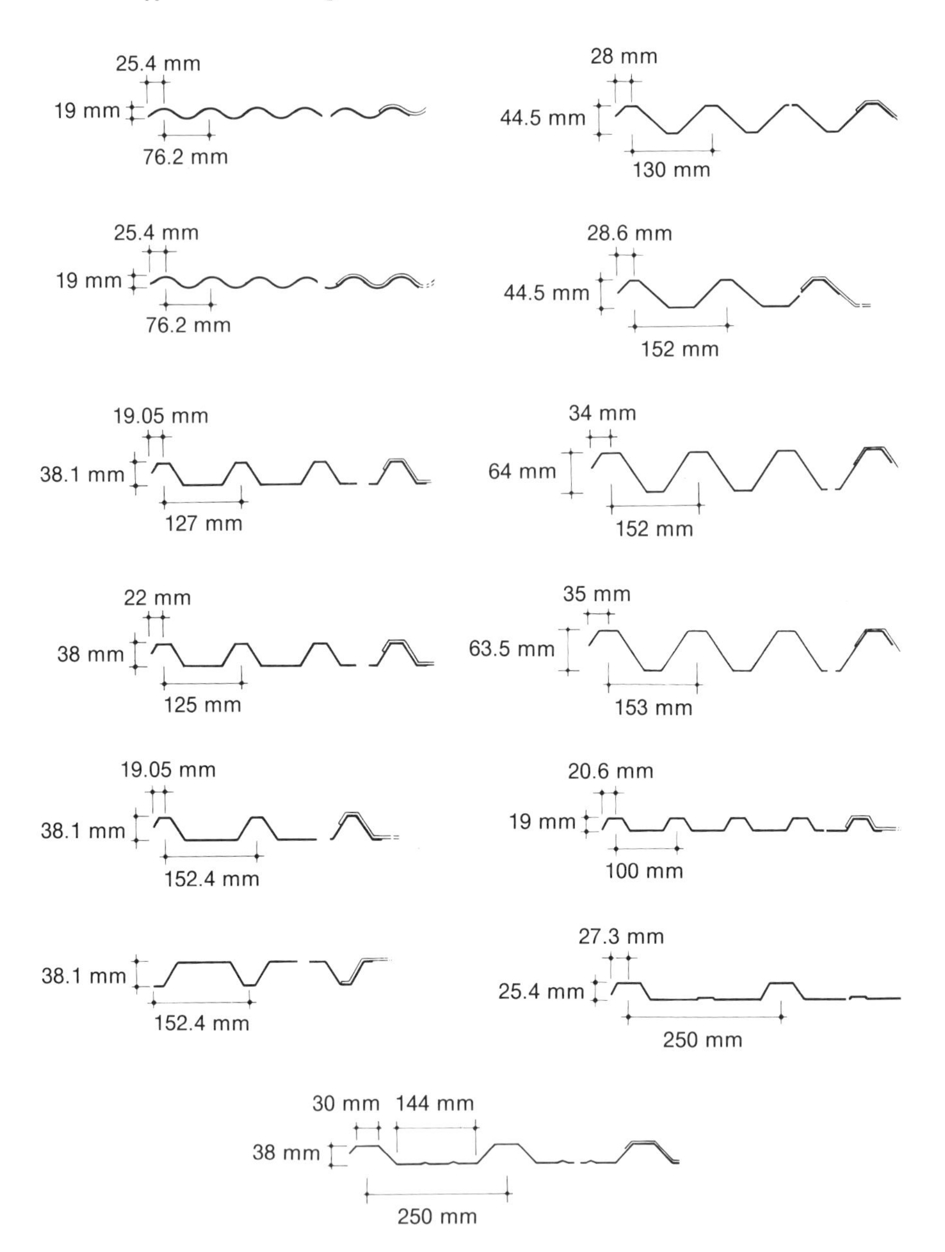

Figure 12.1 **Typical profiles available in aluminium sheet. Generally such material is available in plain mill, stucco-embossed, pre-painted or post-painted finishes in a wide range of colours (courtesy British Alcan Building Products)**

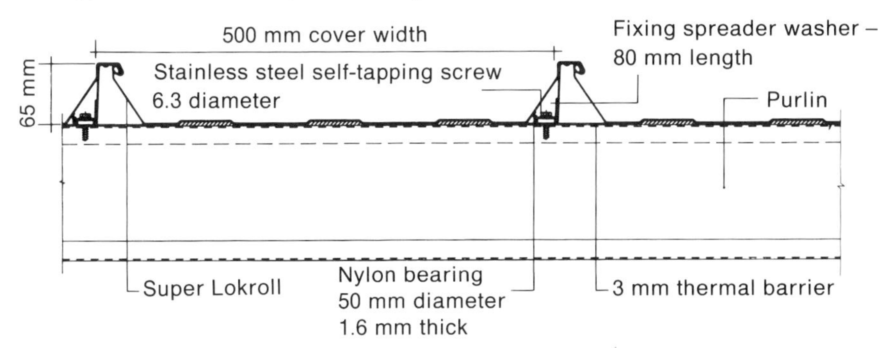

Figure 12.2 A secret-fix roofing system ideal for roof pitches down to 1.5 (courtesy British Alcan Building Products)

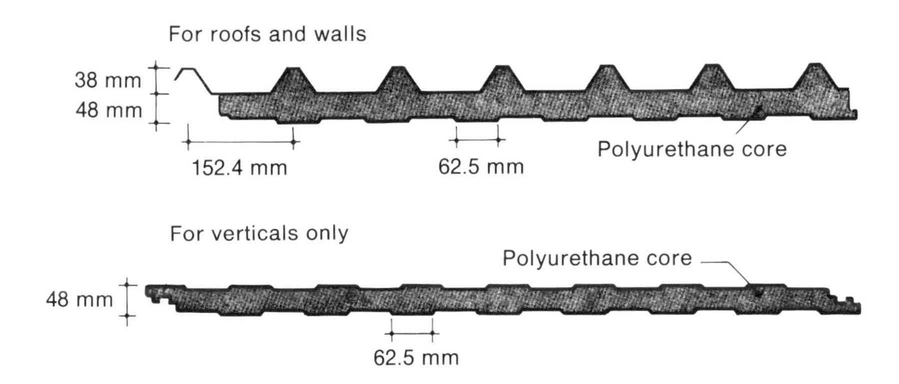

Figure 12.3 Typical composite panels incorporating an inner and outer skin of rigid profiled aluminium sheet with a factory-injected core of rigid closed cell polyurethane foam. These panels typically have 'U' values of less than 0.40 W/m²degC

sheet. This gives additional design flexibility and enables roofs to be specified with a blend of curved and straight sheets of the same profile.

An exciting alternative to both of these production methods is offered by superplastic forming. This technique, applicable to specialized alloys only, offers greater depths of forming and allows more decorative patterns and shapes to be incorporated than are possible with standard forming methods.

Flashings are available in various thicknesses. The versatility of fabrication with aluminium sheet means that virtually any detail required can be accommodated. Rigid flashings in a range of finishes to match the roofing finish – mill finish, stucco, alocrom or painted – are complemented by soft aluminium high-purity flashing available in rolls for on-site forming

This CEGB building in Gloucestershire is roofed with blue pre-coated aluminium sheeting with distinctive rounded eaves. Side-wall cladding is finished in a contrasting white

The North Terminal at Gatwick Airport makes extensive use of aluminium cladding to provide a functional, adaptable building with an impressive facade. The terminal incorporates many interior uses of aluminium also

Close-up of some of the many ribbed panels formed from super-formed aluminium sheeting used at the North Terminal, Gatwick Airport

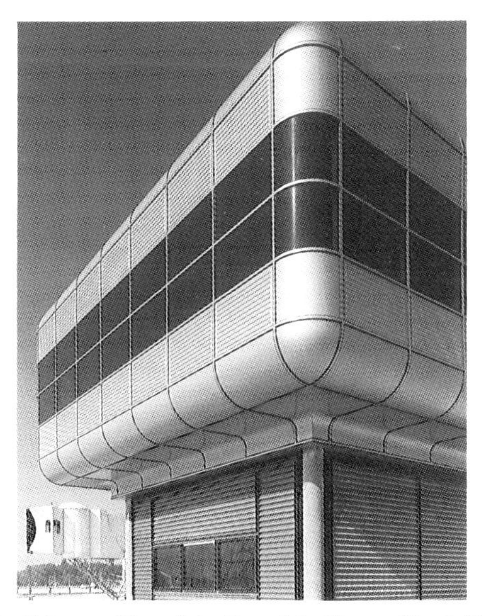

Cladding panels for the Western Pier of the Gatwick Airport North Terminal are made from super-formed aluminium panels backed with a polystyrene core and an aluminium sheet lining panel

Table 12.1 Typical alloy compositions for building sheet

Application	BS Specification 1470	Temper	ISO Designation
Building sheet	3105	H6, H8	Al–Mn–Mg
	3103	H8	Al–Mn1
Flashings (rigid)	3105	H4, H6	Al–Mn–Mg
	3103	H4, H6	Al–Mn1
Flashings (ductile)	1050A	O	Al 99.5

operations (Table 12.1). Step flashings, soakers, dormers, canopy tops and cheeks, valley flashings, copings and linings can all be produced this way.

Ductile and easy to work, such flashing offers the advantages of being cheaper and safer than lead, and more durable than zinc.

Vapour barriers

For some buildings single-skin roofing and wall cladding is adequate weather protection. During wintertime or cold evenings condensation droplets may occur on the underside of single skins. Where this is unacceptable it can be remedied by the use of an anti-condensation paint treatment.

In the UK, however, Building Regulations requirements regarding thermal transmission through the fabric of a building generally call for the use of additional insulation. Such insulation can be provided by the use of composite panel constructions involving an inner and outer layer of aluminium with a core of insulation material such as polyurethane foam. With insulated panels of this kind the internal surfaces, in normal conditions, are kept warm enough to avoid condensation.

As an alternative to composite panels, separate insulation panels may be fixed on the underside of single-skin cladding. Such panels frequently include aluminium foil in the construction as a means of providing a good weatherproof barrier and protective skin for any underlying moisture-absorbent insulant material. Insulation panels of this construction are also extensively used as linings in traditional brick and tile constructions.

Table 12.2 shows a comparison of various vapour barrier materials.

Sound insulation

There are no specific Building Regulations requirements covering sound insulation in industrial and commercial buildings. There are, however,

An aluminium foil cored vapour barrier was used in the reroofing of the industrial unit at Eastbourne General Hospital (courtesy D. Anderson)

Foil-faced insulation panels on the walls of this home provide a snug heat-insulating outer envelope (courtesy Celotex Ltd)

Double-faced aluminium foil foam boarding being fitted to the roof of a new home in order to give a high thermal resistant insulation covering (courtesy Celotex)

Table 12.2 Comparison of various vapour barrier materials*

Description	Material	Approximate vapour resistance MegaNewtons seconds/gram
High efficiency barrier	Aluminium foil	up to 4000
	Insulex 370	1386
Medium efficiency barrier	Polythene sheeting	250
Low grade vapour check	Zinc oil paint on wood	50
	Glass oil paint on plaster	30

*Figures courtesy of British Sisalcraft *Technical Data Sheet.*

occasions when local authority or other specifiers demand specific levels of sound insulation.

Aluminium cladding on its own does not offer a significant barrier to noise. The use of underlays consisting of high-density mineral or glass fibre mats is very beneficial and not only reduces noise but provides effective heat insulation too.

Fire resistance

Non-combustible materials do not assist in spreading a fire. Aluminium does not burn, and mill finish aluminium is classified as non-combustible, so that in consequence it is rated 'Class O' as defined in Building Regulations.

An aluminium rainscreen was successfully applied in 1989 to the badly leaking Millbank House, Southampton. This 16-storey block of flats was erected in the 1960s. Ribbed aluminium panels and high-performance, aluminium-clad timber windows fitted by aluminium mullions to the existing structure have revitalized the building. Aluminium balustrading was also added in the renovation

In many cases walls and roofs can be designated as 'unprotected areas' in accordance with the Governing Regulations. This means that there is no requirement for a period of fire resistance (see BS476 Part 8 for details). In such cases it may be considered an advantage if the roof or wall concerned burns through relatively quickly at the heart of the fire providing an escape vent for heat and smoke to the outside atmosphere. The lower melting point of aluminium compared with steel may be considered an advantage in this respect.

Rainscreen cladding

Rainscreen and overcladding systems are becoming increasingly popular for the refurbishment of industrial and commercial buildings.

The principle is to provide a back-ventilated and drained flat panel cladding envelope that is erected over or around existing structures to protect those structures from the effects of changing atmospheric conditions. The light weight of the aluminium system imposes minimal strain on the supporting walls and provides an external skin that weatherproofs aged brick, stone

Millbank House before its restoration

and cement work, while at the same time giving the building an attractive facelift to its external appearance.

The object of this cladding is not to provide a hermetic skin, but a 'breathing' structure. A ventilated cavity is formed between the existing fabric and the additional cladding. This combats condensation, allows drainage and evaporation of moisture and permits the original fabric to release any entrapped moisture. Figure 12.4 illustrates overcladding principles for various situations.

Structural framing and cladding elements are produced using both aluminium sheet and custom-designed extrusions. A variety of fixing methods have been developed to suit different requirements. The methods may be described as:

Face fix using fastenings such as rivets visible from the exterior (Figure 12.5(a)).

Semi-secret fix using fastenings such as rivets that locate in rebated joints, subsequently disguised by cover strips (Figure 12.5(b)).

Secret fix a variety of clip attachments and concealed fittings exists (Figure 12.5(c)).

Cassette fix panels are mounted using slots or keyhole apertures onto appropriate fixings provided by the framework of the added substructure (Figure 12.5(d)).

The overcladding system lends itself to attractive colour coordination using either anodized and/or powder painted components.

This 11-storey housing block at Chertsey Crescent, Surrey, is another 1960s building to have been renovated 20 years later using an aluminium rainscreen cladding system. Water ingress, rotting windows, condensation and concrete degradation were all problems remedied by installing the overcladding

Detail of the fixing of the aluminium outer cladding to the original fabric

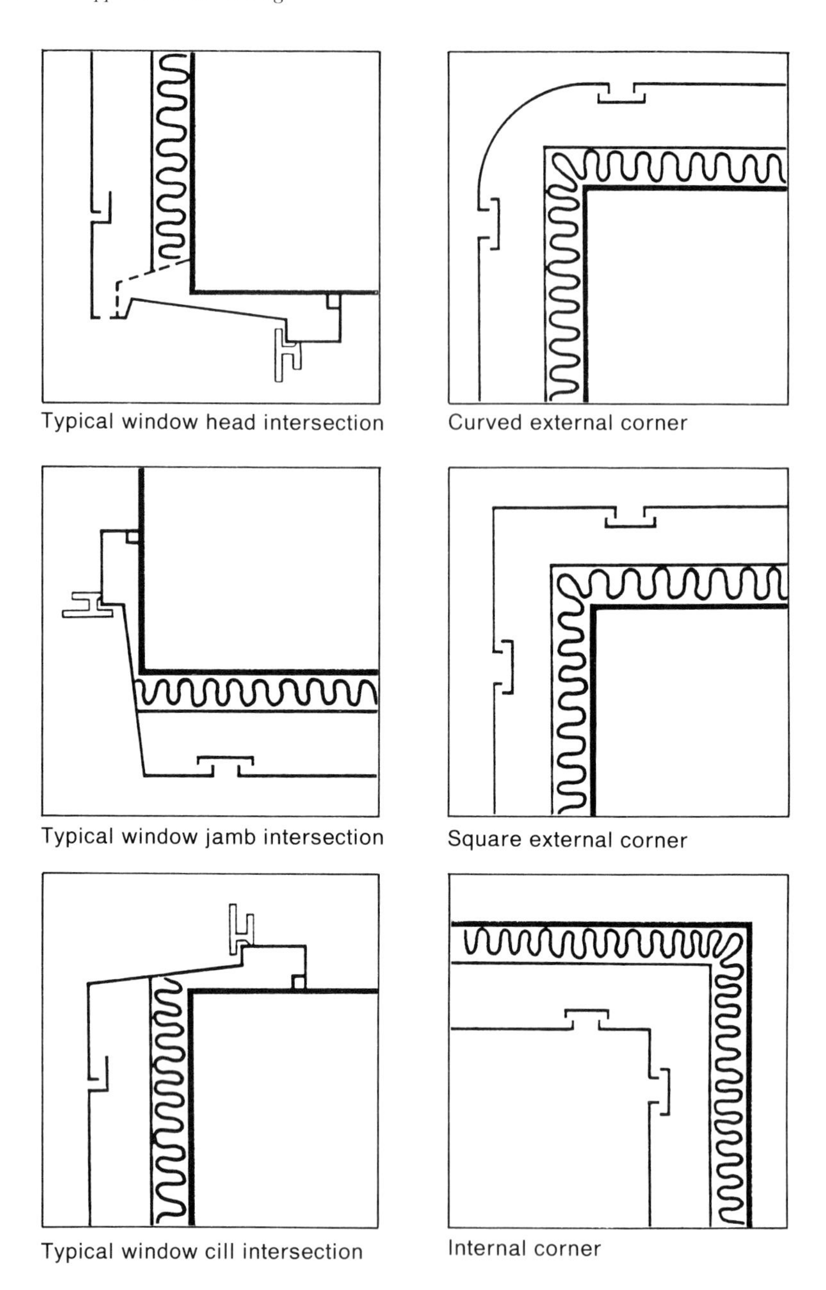

Figure 12.4 Overcladding principle for various situations

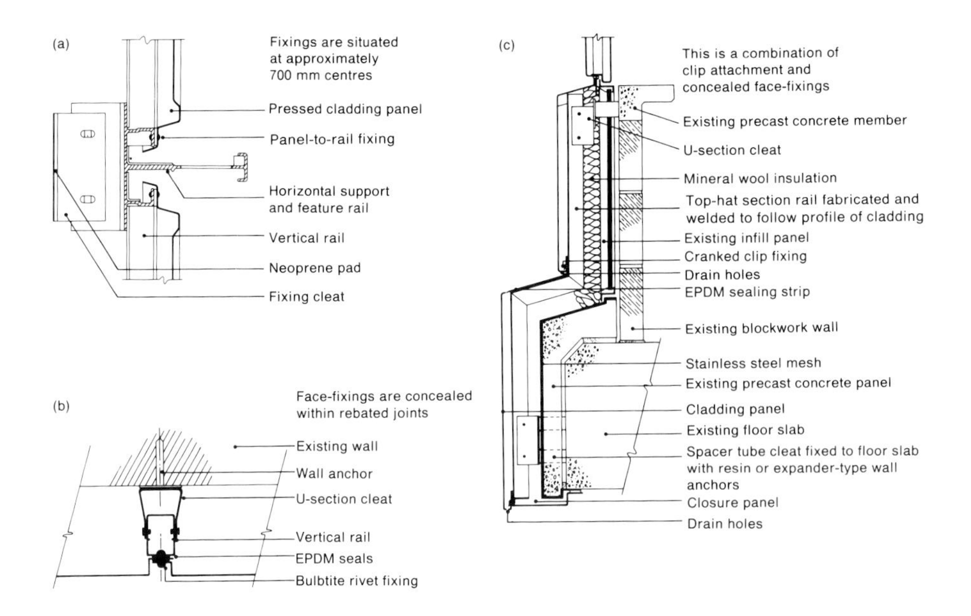

(d) Panels are mounted via slots or key hole apertures onto appropriate fixings situated in the substructure

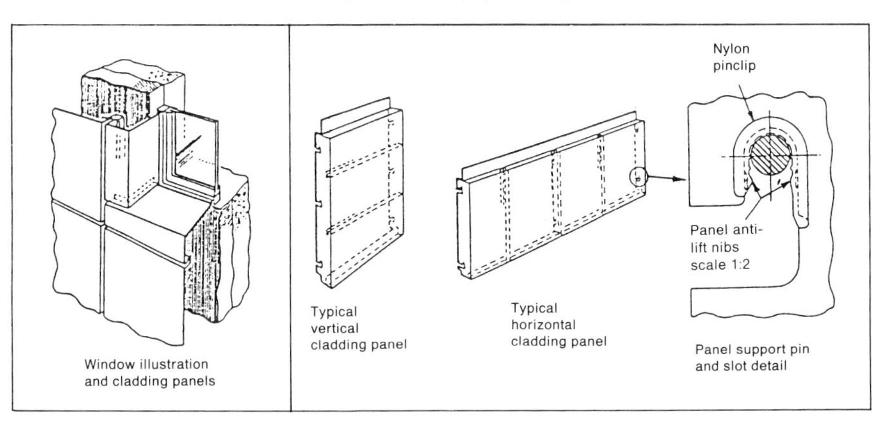

Figure 12.5 Typical overcladding fixing methods: (a) face fix method, vertical section; (b) semi-secret method, horizontal section; (c) secret fix method, vertical section; (d) cassette fix method, perspective view (all courtesy Baco Contracts)

13 Conservatories

One of the success stories of the 1980s was the resurgence in popularity of the conservatory, based largely on special designs and systems of aluminium extrusions. Whereas the standard garden greenhouse is made of mill-finish aluminium, the standard conservatory mainly embodies white-painted aluminium. Many designs utilize single-glazing only, but a number of more sophisticated 'up market' systems now incorporate both double-glazing and thermal-break extrusions. Detailing of design is facilitated through the ease of producing custom-made sections by extrusion and by the relative cheapness of making the required dies. As a result, inbuilt features now include drainage channels, gutters, concealed wiring tracks and weatherseal gasket fixings.

The strength of aluminium alloy is important in conservatory design, particularly for roof members where it is essential to have properly stressed sections that will withstand snow and wind loadings.

Designs range from simple domestic 'lean-to' styles to sophisticated 'Victoriana' designs and shapes. White painted finishes have become the accepted aluminium standard. But other colours are available and brown is a popular alternative.

Both the domestic and commercial markets are important for conservatories. On the domestic scene the ease of erection of these 'add-on' rooms coupled with freedom from involvement with planning permission has made conservatories an ideal way of providing an attractive house extension. Commercially hotels and other establishments have benefited from the same advantages.

A rectangular 'lean-to' style single-glazed aluminium conservatory

A double-glazed 'Victoriana' style conservatory incorporating contrasting hardwood surrounds

Interior view of a double-glazed aluminium framed conservatory

A typical guttering detail for an aluminium conservatory

Mullion and transom fixing detail on an aluminium conservatory

An unusual conservatory application providing extra living space to an upstairs room (courtesy Kaye Aluminium)

This rotating pyramid glasshouse provides a novel approach to undercover gardening in confined spaces. All components, base, glazing bars and cone are constructed of aluminium

A room extension with an octagonal conservatory 'Victoriana' style feature. Powder painted thermally-broken extrusions were used throughout

14 Patent glazing

In essence, patent glazing is a method of providing glass-panelled structures ranging from a simple domestic glass-house to the largest of covered shopping malls and centres, factories and power stations.

Perhaps the most important patent glazing application in the UK, and possibly in the world, is the St Enoch Square shopping centre in Glasgow. A unique glass envelope containing 28 000 sq m of vertical and roof glazing covers the entire shopping and leisure centre, providing a completely controlled internal environment. This imaginative structure, completed in 1989, is a tribute both to the architects and the engineering and building skills provided by the British companies involved.

Traditionally, glazing bars provide a two-edge support for the glazed cladding, which may be either single or double panels. The glass is held in place by clips, generally applied in continuous lengths to produce an even pressure along the glass edges. Figure 14.1 shows bar profiles for single and double glazing.

A typical standard glazing bar is a bulb tee section, providing the combination of stiffness and lightness, with incorporated double drainage channels and built-in grooves to incorporate the spring glazing wings.

Extruded sections for glazing bars, heads, cills and various fittings are usually extruded in alloy 6063T6. The spring glazing wings are cold roll formed from sheet alloy 3103. Metal flashings are usually formed from alloy 1200 or 5005, the latter being specifically specified where an anodized finish is required.

Many patent glazing applications are acceptable in mill-finish

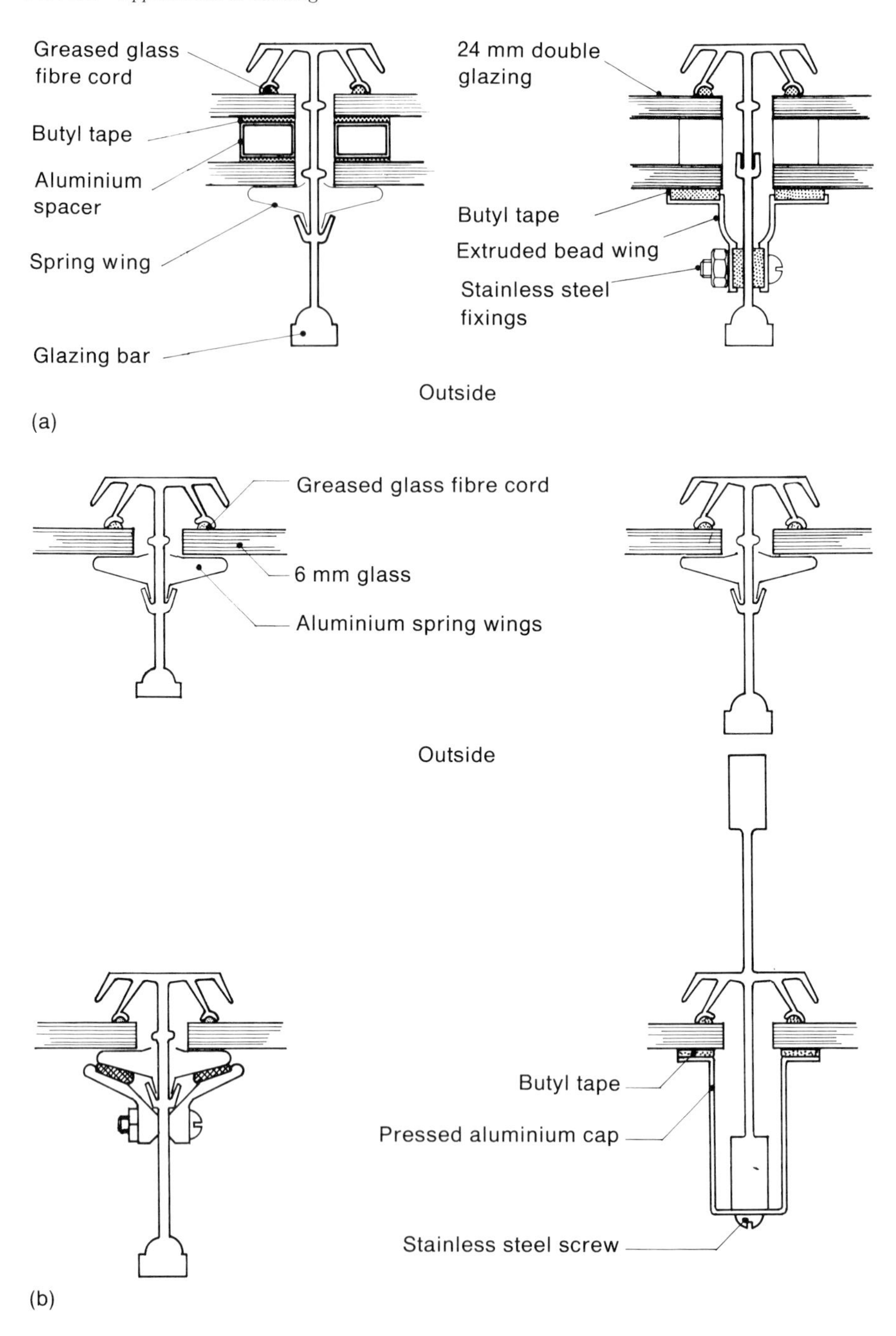

Figure 14.1 Patent glazing bar profiles: (a) double glazed; (b) single glazed

The glass cover for the St Enoch Square shopping and leisure precinct in Glasgow – the largest glasshouse in the world – dominates this aerial view of the city alongside the River Clyde. The 28 000 m^2 of vertical and roof glazing incorporates 17 glazed pyramids. Naturally anodized aluminium is used for interior components while the external surfaces are powder coated. The project was completed in 1989 (courtesy Mellowes PPG Ltd)

An interior view of some of the patent glazing at St Enoch Square

Aluminium glazing contributes to the impressive atrium at the Aztec Centre, near Bristol

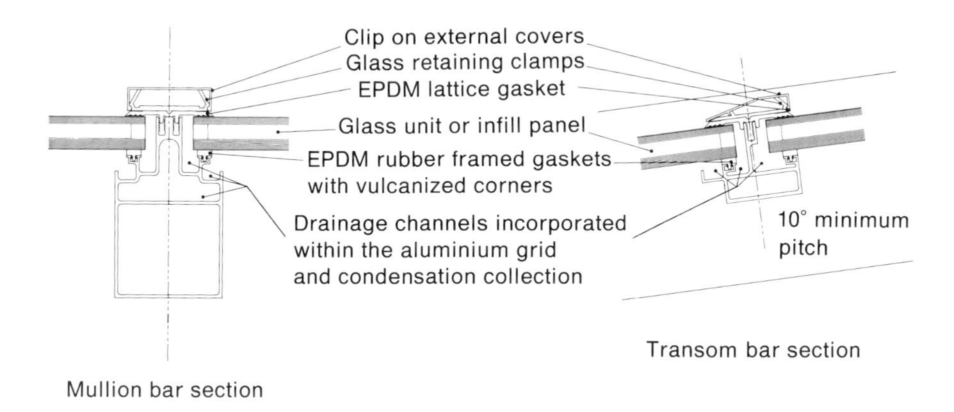

Figure 14.2 A high-performance roof glazing system design primarily for shallow pitch situations. It is fully drained and thermally broken (courtesy Heywood Glazing Systems)

aluminium, but for those applications where decorative considerations demand a more pleasing appearance, then either anodizing or painting may be specified. Patent glazing may be used for vertical surfaces and for roofs with pitches down to 5° (depending upon design).

For applications demanding higher performance than is obtainable with standard glazing bar sections, other specially designed sections are available. Figure 14.2 illustrates a double-hollow section that gives high strength and incorporates various drainage channels.

The monorail transport system at the Merry Hill, West Midlands, shopping complex has tubular stations embodying aluminium patent glazing and curved infill panels

An interior view of one of the monorail stations at Merry Hill, taken during construction, showing the glazing bars, and the external aluminium skin of some of the cladding panels. The curved panels were of varying constructions but typically comprised aluminium sheet/insulation core/aluminium sheet.

A series of glazed pitched roofs incorporating aluminium patent glazing and curtain walling form the main entrance to a Safeways supermarket at St Helens, Lancashire

Aluminium glazing continues to be the first choice for commercial and domestic glasshouses. Invariably in mill-finish metal, aluminium glasshouses give years of trouble-free service (courtesy Cambridge Glasshouse)

A dramatic view of one of the two glazed domes at Whiteleys department store, Queensway, London, after being reglazed with aluminium patent glazing (courtesy Ruberoid)

Manchester's old Central Station was given a new lease of life when an imaginative refurbishment scheme converted the Victorian barrel-vaulted building into an exhibition centre. Extensive use of aluminium was made for the patent glazing, all with a black polyester powder finish, and the roofing, finished with a PVF2 granite-grey long-life coating

15 Space frames

In the creation of structural frameworks, aluminium competes directly with steel. Because of the differences in physical characteristics between the two metals, aluminium can only compete economically and effectively where full advantage is taken of the metal's excellent strength-to-weight ratio, or where light weight is a critical factor, as in off-shore oil rig structures.

The characteristics of aluminium have been put to impressive effect with the development of spectacular space-frame structures of which Triodetic is notable. Triodetic structures consist of an assembly of aluminium tubular members coupled together using patented joints. These can be joined together to form two-way, double-layer, flat space frame grids with clear spans of up to 40 m, three-way grids in a similar span range, barrel vaults up to 50 m span, various composite structures such as A-frame and V-frame trusses, domes up to 70 m diameter and pyramids up to 50 m clear span (Figure 15.1(a)–(c)).

Adaptations of this space-frame grid can be used to construct free-form structures of complex geometry such as hyperbolic paraboloid shells.

The Triodetic space frame construction has been used in recent years for exciting projects on a world-wide basis, including the giant domes of Shah Alam state mosque at Selangor, Malaysia, a free-standing double-curvature feature roof for a marina at Niagara Falls, Canada, and by contrast, a glazed pyramid roof over a leisure centre at Romford, Essex.

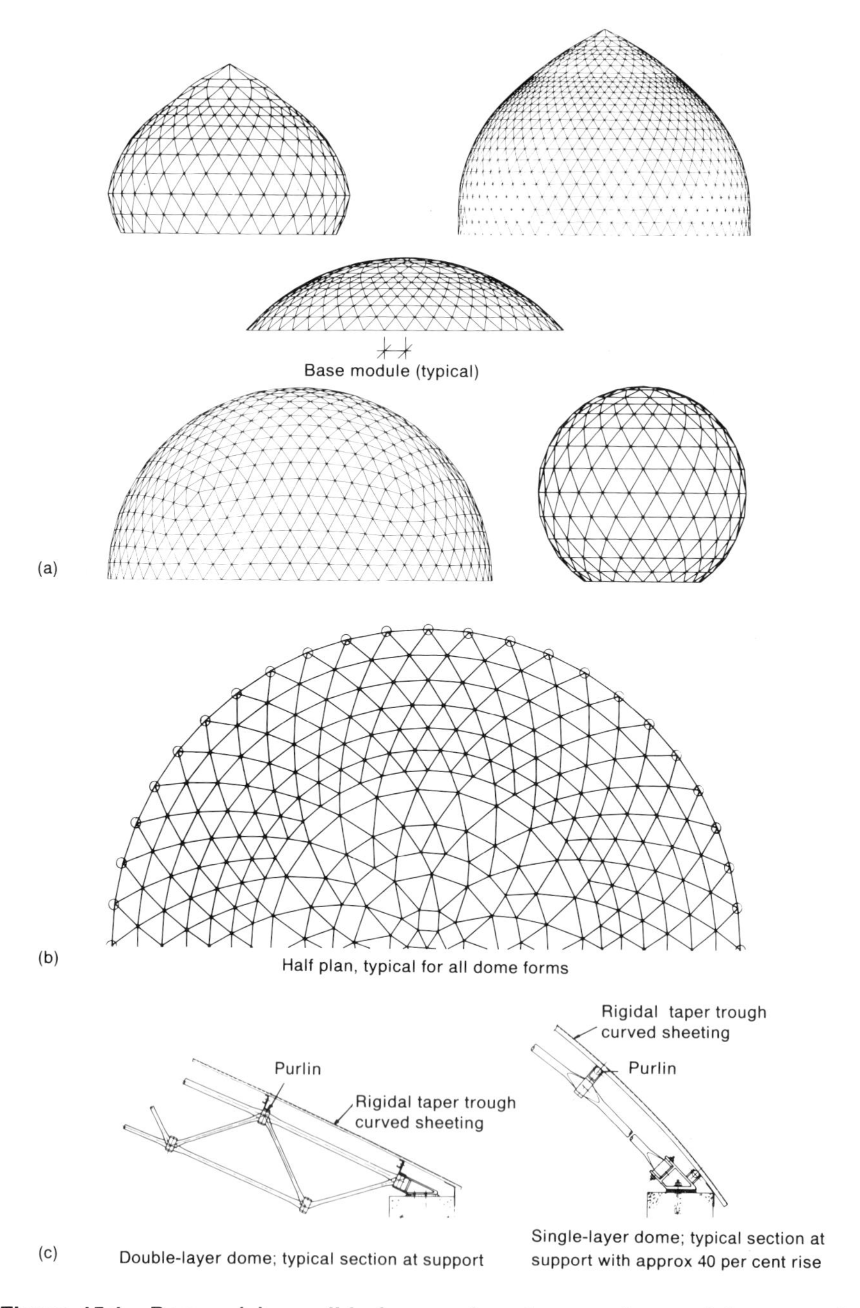

Figure 15.1 Domes (a) possible forms: almost every shape of dome can be constructed using the Triodetic system: (b) half plan; (c) double-layer and single-layer domes (courtesy Baco Contracts)

A space frame constructed from a patented tubular system (courtesy Spherobat)

Space-frame structures not only lend themselves to innovative design but also offer easily handled and erected components that enormously simplify site handling and installation.

Fitting infill panels to a Triodetic aluminium space frame (courtesy Baco Contracts)

Erecting a space-frame structure for a mosque in SE Asia

The splendid dome of the Selangor mosque is built with a Triodetic space frame for lightness, strength and ease of construction (courtesy Baco Contracts)

The Triodetic space frame covering a swimming pool in Basildon, Essex, gives a light, airy impression to the roof structure

This multi-curvature, free-standing Triodetic space structure makes an unusual centrepiece for this dolphin pool in Canada

16 Highway furniture

Balustrades

Aluminium in cast, extruded and plate forms is being increasingly used for a diverse range of exterior balustrade and parapet applications, particularly on bridges. There are a number of reasons for this. The metal has a high strength-to-weight ratio, it is very durable and may be safely left for many years without painting or applying a protective coating, and it has a high energy-absorption coefficient. These advantages combine to make aluminium an acceptable and frequently preferred alternative to steel for many locations where safety barriers, either for pedestrians or traffic, are required.

The three basic alloy materials used in parapet construction are BS1474 6082T6 for extrusions, BS1470 6082T6 for plate and BS1490 LM6M for castings. Mill finish, or as cast finish, is acceptable.

Where welded posts and joints are employed it is important to use MIG welding techniques with the choice of filler wire selected to suit the alloys being welded. Where extrusions or plate are welded to a casting, the filler wire choice should be BS1475 4043A; all welds between wrought components should be completed with BS1475, 4043A or 5056A. Various ingenious patented systems of 'secret' fixings have been developed, and at least one hand-rail system incorporates integral lighting strips by using specially designed extrusions. Bolts and washers should be of stainless steel.

Aluminium parapet railing with a fitted wire mesh pedestrian safety barrier

A square tubular design of aluminium bridge railing

Heavy-duty bridge parapet across a Scottish loch (courtesy Baco Contracts)

Another major highway product where aluminium offers advantages, but where so far it has made only modest inroads in the UK, is lighting columns. Such columns could be the answer to the serious problem and expense of maintaining installed steel columns on many of Britain's motorways and arterial roads.

Lighting columns

Lighting columns in aluminium have been successfully used in shopping centres, car parks and private roads for many years. In main road applications where taller columns are required there has been a lower acceptance because of aluminium's higher initial cost, compared with steel and concrete. It is when the additional costs of maintaining steel columns in particular are included in the cost equation that aluminium looks more cost-effective.

The improved safety aspect of aluminium columns is another potential advantage. Experimental work in the UK, and many years of field experience in the USA, indicate that personal injuries caused when a vehicle is in collision with an aluminium column are less serious than when a vehicle collides with either a steel or a concrete post. Experience

Class 'A' aluminium lighting columns on a busy highway. The mill-finish columns provide a long, virtually maintenance-free life

Fabricating the trunks of aluminium lighting columns (courtesy Thorn Electrical)

Direction signs and road signs are made from both aluminium sheet and extruded aluminium plank sections. The light weight and good atmospheric durability are key features

suggests that the higher energy-absorbing characteristics of aluminium are a contributory factor in reducing road accident severity.

Even higher levels of personal safety are achievable with aluminium columns fitted with breakaway bases. The use of such columns is commonplace in the USA and good results are reported.

Both extrusions and sheet are used in lighting column construction. Many of the larger columns are constructed from formed and welded sheet using alloy BS1470 5251.

17 Shopfronts and ground-floor treatment

In the highly competitive environment of the shopping high street, an attractive frontage and good merchandise visibility are essential for survival.

To achieve this combination aluminium shop fronts and ground-floor treatment have been installed virtually to the exclusion of all other materials. The slim lines obtained by using specially designed suites of aluminium extrusions provide attractive appearance, strength and maximum uninterrupted display area, together with all the advantages of aluminium's durability and requirement for minimum maintenance.

Window frames, entrance doors, canopies and porches are available not only in many modern designs but also in a simulated Victorian appearance, obtained by the use of special decorative features. Anodized finishes are very popular, with specifiers choosing widely from the full range of colours now available through the appropriate selection of anodizing process.

Perhaps more importantly, however, the much broader colour range offered by powder coating has been decisive in the architect's mind. Using a powder paint finish the shop-owner can select a coloured frontage that is distinctive, unique and which reflects the corporate identity and colour scheme of the company.

Figure 17.1 illustrates typical shop-front constructions.

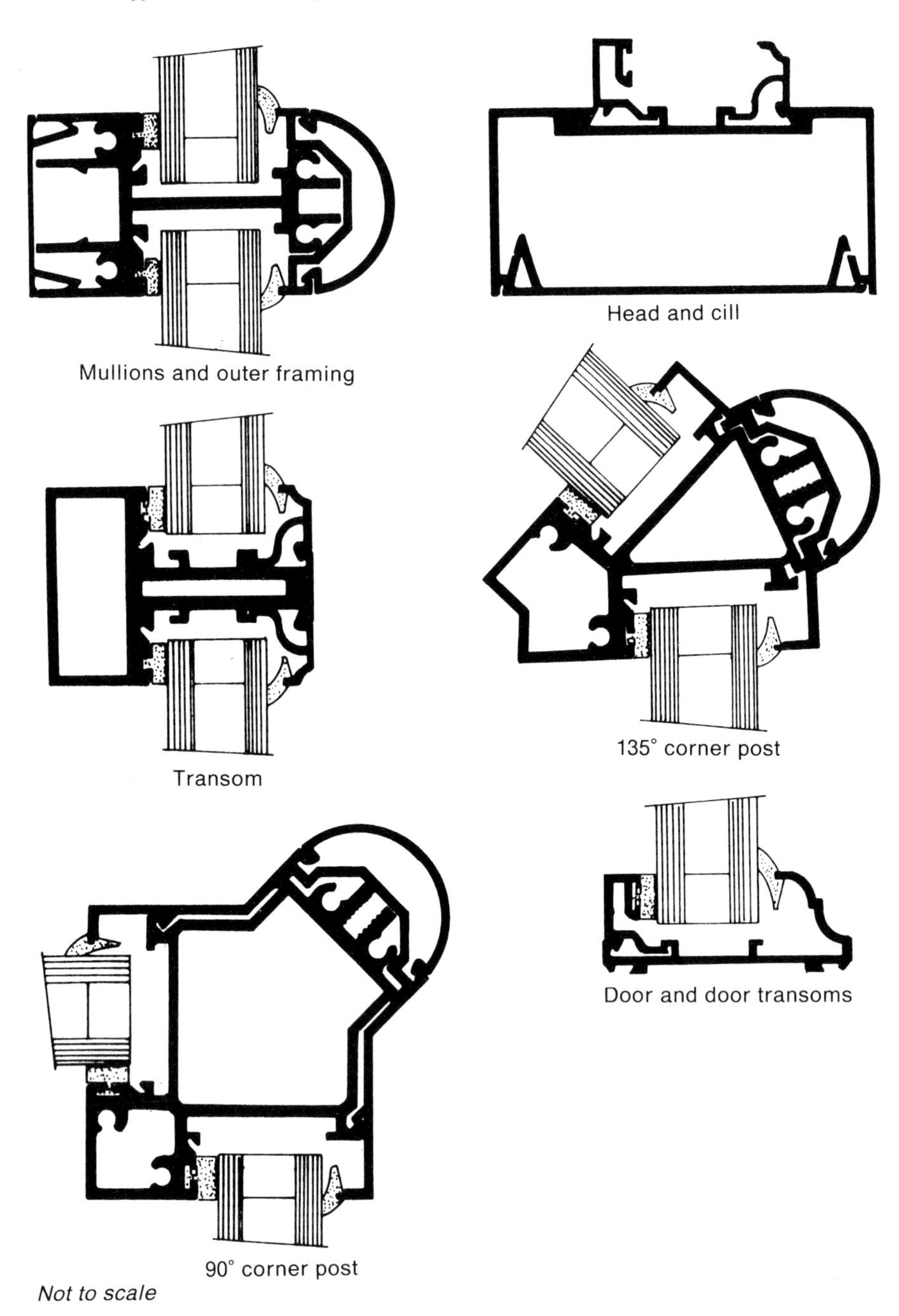

Figure 17.1 Typical details of shop front construction (courtesy Heywood Doors Shopfronts)

18 Rainwater systems

Aluminium rainwater systems are becoming increasingly popular, and cater for all types of domestic, commercial and industrial building, either of traditional or modern design. Competing with steel and plastics, aluminium rainwater goods offer the advantages of light weight, ease of erection, and above all, durability.

Plain finish products are available, but much more in vogue are the polyester paint powder-coated finishes that are available and which can complement or contrast with brickwork, windows and roofing as required. The durability of both painted and unpainted aluminium rainwater goods is excellent. Manufacturers' claims for such products refer to a minimum maintenance-free life of 40 years in rural and suburban conditions, and of 25 years in industrial and marine environments.

Cast, rolled and extruded forms of aluminium are all used. Generally cast gutters and fittings are made from LM2 and LM6 specifications; rolled gutters and other sheet-based items are generally 3103 or 5251, while extruded components are mainly made of 6063T6.

Guttering styles

Styles are available to suit all kinds of properties from old to new. Traditional half round gutters are widely specified, but also Victorian Ogee gutters and box gutters are available. Rainwater downpipes may be either round, square or rectangular.

Aluminium guttering and downpipe made on site from roll-formed pre-painted aluminium coil. Note the pleat formation that enables curves to be included in the downpipe

Roll-forming aluminium guttering on-site from the back of a van. The technique enables guttering to be quickly and easily erected with a minimum of joints as lengths can be rolled to the exact size required

A 'flush-fit' cast aluminium downpipe with no climbing aid projections. Specially designed as a 'no-climb' pipe (courtesy Alumasc)

An attractive 'Victorian' style cast aluminium ornamental hopper head (courtesy Alumasc)

'Seamless' guttering

The benefits of pre-painted aluminium alloy coil are put to good use by manufacturers offering on-site guttermaking and fixing. By forming exact lengths of guttering on site, using a travelling roll-forming machine, it is possible not only to offer a rapid, efficient delivery and fix service, but also to avoid the joints that occur when fixing together standard lengths of guttering.

Compatibility

Direct contact with dissimilar metals or with materials containing lime or cement should be avoided. Any gutter joints should be insulated using a low modulus silicone sealant. A liberal coating of bitumen is recommended as an insulant between, for example, lead flashings in contact with aluminium gutterwork, and between downpipes and any touching cement or brickwork.

The use of aluminium guttering and fixings in conjunction with roofs having copper flashings or cladding is not recommended.

Suitability

Aluminium has built up a solid history of performance for rainwater goods. It was first used in 1945 and since that date its durability has been well proven. Its strength is adequate to withstand snow loadings and also to support ladders.

The advent of powder-coating techniques and of pre-coated aluminium coil has added a colourful new dimension and led to a significant growth in demand. Much growth potential still exists in the UK, as currently pre-coated rainwater products account for just 5 per cent of the total market, whereas in the USA pre-coated aluminium accounts for a dominant 80 per cent.

19 Architectural metalwork

Sculpture

Aluminium is an easily workable material. It can be cast, or shaped by various means ranging from hammering and beating through to spinning, pressing and forging. Its surface texture can be varied by chemical etching, anodizing, scratch brushing, embossing, peening, or polishing and of course it can be painted and lacquered either using opaque or translucent coatings.

One of the world's sculptural landmarks – Gilbert's 'Eros' statue in Piccadilly Circus, London – is a tribute to the use of aluminium. Cast in 1893 it was one of the very first uses of what was then a very new material. Despite being cast from an 'impure' alloy the condition of the statue has remained excellent.

For architectural sculpture the lightness of aluminium is particularly valuable, enabling large and imposing figures to be erected without creating structural overloads on the flooring or walls of a building. Reduced transport and erection costs are an additional benefit.

Various alloys can be used for sculptured objects, and all forms of aluminium, cast and wrought can be employed in a fabrication. Modern welding methods, coupled with the more recent availability of adhesive bonding have widened the designer's scope, enabling large, intricate objects to be easily constructed.

Aluminium sculptures are highly suited for both interior and exterior applications. Ecclesiastical uses have been very popular, with interesting and imaginative applications to be seen in a number of cathedrals. One of the latest in this line of church-style artefacts is the interesting piece constructed in 1990 by designer Michael Joannidis while a student at the Royal College of Arts, London. Made of polished aluminium plate and adorned with cast aluminium lamp brackets supporting glass orbs filled with coloured paraffin lights, this imposing modern design, entitled *Altar Lights*, makes an impressive use of aluminium's properties.

Hardware and fittings

The growth in use of aluminium doors and windows has naturally led to a demand for matching hardware for items such as handles, bolts, stays and letter flaps. But the use of aluminium for architectural hardware spreads very much wider – aluminium hardware is used to complement products of other materials. Aluminium door furniture, and indeed furniture trim of many kinds, either cast or cut from extruded sections, has become over the past 30 years as accepted and as popular as brass fittings had been in earlier times.

Both anodized and lacquered finishes are used to enhance appearance. Lacquered finishes are less permanent than anodized ones and are more susceptible to wear and abrasion. Anodizing, which may be applied on

Aluminium door closures

A cast aluminium door handle with black powder paint finish

top of plain or textured aluminium, need not be as thick for interior applications as the 25 μm recommended for exterior applications. The lowest recommended grade for interior applications is 5 μm (AA5), but this should only be used where very little wear is anticipated. A minimum thickness of 10 μm, AA10, is generally recommended.

Anodized finishes are applied in natural silver, and in a range of colour options of which golds, bronzes and black are the most popular.

As in many other areas of aluminium usage, powder coating has extended the range of colour finishes and it is now possible to get hardware that is colour coordinated with window and door frames.

20 Interior applications

Partitions and ceilings

Aluminium partitioning and screening, built up on systems of extruded aluminium frames, is ideal for subdividing room space. Either permanent fix or demountable structures are available. Slim aluminium frames combine lightness and elegance with strength and functionality. Hollow extrusions or clip-together solid sections can be used which incorporate hidden channels for the passage of electrical wiring and communications cabling.

Aluminium framed partitioning can be either solid-filled with a variety of panel materials or glazed, as appropriate.

Suspended, or false ceilings, can be constructed along similar lines to vertical partitioning, using a 'frame-and-fill' method with aluminium extrusions again providing the framework. Alternatively an entire 'open' ceiling structure can be assembled using either aluminium strip or extrusions. The use of pre-painted or lacquered aluminium coil is particularly appropriate for ceilings of this type.

Ducting and trunking systems

Aluminium is widely used for a variety of industrial and commercial ducting applications. Heating and ventilating systems make good use of

Bronze anodized and glazed partitioning in an office reception (courtesy Finalex)

An aluminium suspended ceiling constructed from pre-painted aluminium coil (courtesy Alcan Duralcote). The entrance door is silver anodized aluminium and extrusions are used in the showcases

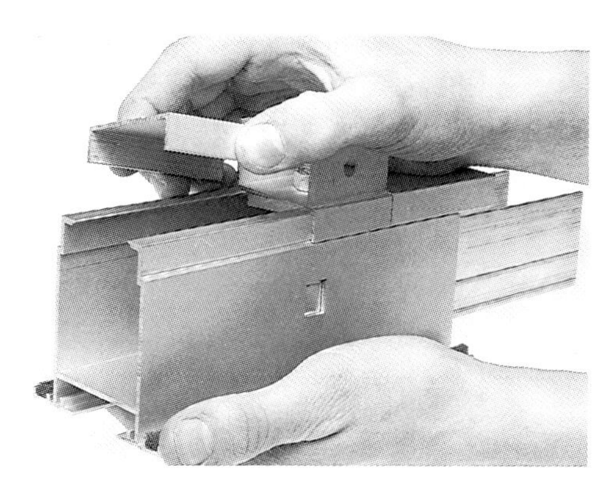

Components of a suite of aluminium extrusions designed for a suspended ceiling framework. The close tolerances obtainable on extrusion enable interlocking and sliding fits to be readily obtained

This vertical lighting column hides an office management system for wiring and cabling. The column is built up from a series of interlocking sections and the smooth exterior contours contrast with the complex detailing contained within (courtesy Supapole 7)

An 'opened-up' view of the sections that comprise the Supapole 7 lighting column

aluminium sheet for its light weight and durability and for its ease of handling, workability and installation.

In modern office environments suites of specially designed extrusions are finding favour as a means of incorporating trunking routes for electrical and communications cables into a planned system. By this means cabling is conveniently concealed so that cables and flexes trailing across floors and desks are avoided, along with the safety hazard that inevitably accompany haphazard cabling.

Grilles

Expanded mesh made from sheet, first perforated and then stretched, has been popular for many years as a decorative grille material. More recently a more robust type of lattice grille made from punched and stretched extrusions, has found applications for decorative partitioning, window security screens and protective, safety grilles for bars and counters.

Others

There are too many interior applications of aluminium, both actual and potential, within a building to discuss in this publication. We can just mention, however, suspended ceilings, cooker hoods, stair tread and nosing, carpet edging, furniture, shelving, radiators, curtain rails, lighting tracks, kick plates, balustrades, panelling, insulation panels and decorative trim.

Extruded and anodized aluminium lighting track was fitted throughout the Trustees Savings Bank offices in Edinburgh

The interior of Gatwick Airport North Terminal uses large quantities of aluminium sheet for panelling

An entrance foyer fitted with bronze anodized panelling, specially selected for its hardwearing properties as well as attractive finish

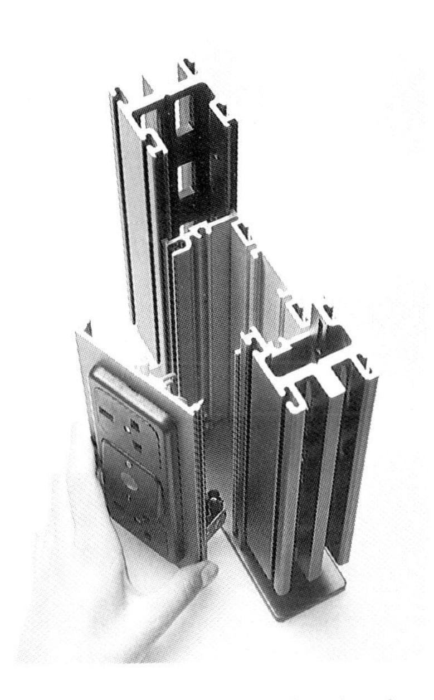

Clip fits, sliding fits and screw location channels are just some of the features built into the extrusions illustrated. They form part of the K-Tek high-tech work centre which is an all-aluminium construction. Designed for computer room and laboratory use the system is extremely rugged and presents clean, modern lines free from clutter and cables

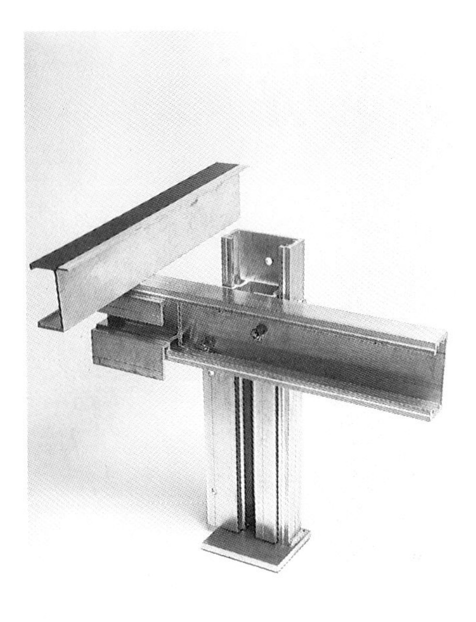

Part of an extruded aluminium supporting structure for a demountable and easily erected portable stage

The vertical supports of this bookshelf system are made from anodized extruded channel section. The horizontal supports that lock into retaining slots on the uprights are machined from a 'T' bar section

This bathroom shower unit is made of anodized aluminium extrusions and corner castings. The excellent durability of aluminium under steamy, moist conditions provides a long-lasting, easy-to-clean unit

This experimental solar heat collection panel makes good use of aluminium's heat conductivity. It is an application not yet exploited in the UK

21 Looking ahead

The conclusion of the first century of aluminium is an appropriate time to take stock of successes and failures and to look ahead towards the next 100 years.

There can be no doubt that aluminium has, in a relatively short space of time made a tremendous impact on modern life. Take away aluminium, and the clock would turn back many years, as many products ranging from aeroplanes to printing plates, and from electric cable sheathing to food packaging rely upon the diverse properties of this twentieth century metal. In more recent times, other materials, plastics and composites have been developed. These have replaced aluminium in some applications, but more importantly have complemented it in others, enabling products to be made – such as the thermally insulated aluminium window – that have been such a success in the last decade.

Environmental considerations are now well to the fore when material selection is being considered. Judged on this criterion aluminium stands to gain. It is in plentiful supply, much more so than iron and steel, oil-based plastics and hardwoods. Importantly too, it is recyclable easily and efficiently so that once produced aluminium can be used again and again; and not just in theory either, for recycling already takes place on a large scale turning scrapped or discarded aluminium products, from old engine components to used drinks cans, back into high-grade products.

Whether any new alloys will be developed, other than perhaps variations on a theme, is doubtful; but in today's highly inventive and technological age such a possibility cannot be discounted completely. It is, after all, only in recent times that the aluminium–lithium series has been developed, with its increased elastic modulus and reduced density. Certainly we will see advances in aluminium composite materials such as aluminium–ceramics and these may open up exciting new possibilities, as will further developments in superplastic forming, powder metallurgy and fibre reinforcement.

Improvements in production and processing methods and techniques which have resulted over the years in increasing metal quality and restraining price rises by improving productivity and efficiency are likely to continue, but at an exponentially slower pace. The Hall–Héroult smelting process, for example, the starting point for all metal production, is basically the same now as it was in 1886. Its efficiency has steadily improved and today the production of 1 kilogramme of aluminium requires less than 14 kWh of electricity compared with 25 kWh in 1950.

Improvements are continuing but clearly at a slower pace. The big question is – when will we see a new smelting process? The advent of a new less energy-intensive method would revolutionize attitudes towards aluminium and expand its uses dramatically.

Expansion using today's existing production methods of smelting and converting seems certain to continue, however. Existing applications such as curtain walling, and to a lesser extent windows, will expand and other established uses are likely to grow in line with general economic expansion. Today's designs of buildings ranging from the Gatwick Airport North Terminal with its extensive use of aluminium in many forms to the magnificent Hong Kong and Shanghai Bank in Hong Kong, with its imaginative use of extruded components, are setting the pace for a greater awareness of the possibilities for aluminium as a modern, sympathetic, building material. With greater use being made of aluminium in prestigious commercial buildings such as these, there is likely to be a greater overall usage of aluminium in building, perhaps extending into a far greater use of aluminium insulated cladding for exterior and interior walls, and maybe for roofing too.

The domestic house building industry is a bastion of tradition but in the long-term economic considerations must lead to an extended use of prefabricated, factory-made components instead of the current on-site methods of construction that are used. At this time all of the experience and confidence that has been built up with aluminium over the years will reap rich rewards. In building, the aluminium age may still be yet to come.

Appendix 1 Relevant standards

The following British Standards and Codes of Practice relate to various uses of aluminium in building and to the metal itself. Also itemized are relevant Aluminium Window Association specifications.

Aluminium

BS1470/1987	Specification for wrought aluminium and aluminium alloys for general engineering purposes: *plate, sheet and strip.*
BS1471/1972	Specification for wrought aluminium and aluminium alloys for general engineering purposes: *drawn tube.*
BS1472/1972	Specification for wrought aluminium and aluminium alloys for general engineering purposes: *forging stock and forgings.*
BS1473/1972	Specification for wrought aluminium and aluminium alloys for general engineering purposes: *rivet, bolt and screw stock.*
BS1474/1987	Specification for wrought aluminium and aluminium alloys for general engineering purposes: *bars, tubes and sections.*

BS1475/1972 Specification for wrought aluminium and aluminium alloys for general engineering purposes: *wire*.

BS1490/ Specification for cast aluminium and aluminium alloys for general engineering purposes.

Finishes

BS1615/1987 Method for specifying anodic oxidation coatings on aluminium and its alloys.

BS3987/1974 Specification for anodic oxide coatings on wrought aluminium for external architectural applications.

BS5599/1978 Specification for hard anodic oxide coatings on aluminium for engineering purposes.

BS6496/1984 Specification for powder organic coatings (on aluminium).

BS4842/1984 Specification for liquid organic coatings (on aluminium).

Structures

BS1161/1984 Specification for aluminium alloy sections for structural purposes.

CP118/1969 The structural use of aluminium.

BS8118/1990 Design Code for structural uses of aluminium.

Welding

BS4870 Part 2* TIG or MIG welding of aluminium and its alloys.

BS4871 Part 2* TIG or MIG welding of aluminium and its alloys.

*Soon to be replaced by EN288 and EN287 respectively.

BS4872 Part 2	TIG or MIG welding of aluminium and its alloys.

Performance

CP153 (Windows and Roof Lights)	Part 1/1969: Cleaning and Safety. Part 2/1970: Durability and Maintenance. Part 3/1972: Sound insulation.

Applications

Windows and Doors

BS5286	Specification for aluminium-framed sliding glass doors.
BS4873/1986	Specification for aluminium alloy windows.
AWA	Specification for aluminium alloy doors and side screens for domestic buildings.

Curtain walling

AWA	Specification for aluminium alloy curtain walling.

Cladding and Roofing

BS4868/1972	Specification for profiled aluminium sheet for building.
CP143(i)/1958	Aluminium, corrugated and troughed.
CP143(15)/1986	Aluminium, metric units.
AWA	Specification for flat faced aluminium alloy building panels.

Patent glazing

BS5516/1977	Code of Practice for patent glazing.

Highway furniture

Balustrades:

BS6779 Part 1	Vehicle parapets.

Lighting columns:

BS5649 Part 3	Specification for materials and welding requirements for lighting columns.

Rainwater goods

BS2997/1980	Specification for aluminium rainwater goods.

Suspended ceilings

CP290/1973	Suspended ceilings and linings using metal fixing systems.

Office furniture electrical systems

BS6396/1983	Code of Practice for electrical systems in office furniture and screens.

Appendix 2 Additional reference material

The Properties of Aluminium and its Alloys, Aluminium Federation
Products and Services Guide, Aluminium Window Association
Guide to the Specification of Windows, Aluminium Window Association
Guidance in the handling, care, protection, fixing and maintenance of aluminium windows and doors, Aluminium Window Association
Guidance to specifiers in the use of mastics and sealants on site, Aluminium Window Association

Appendix 3 Contact addresses

Aluminium Federation Aluminium Extruders Association Aluminium Finishing Association Aluminium Rolled Products Manufacturers Association Aluminium Stockholders Association	Broadway House Calthorpe Road Five Ways Birmingham B15 1TN
Aluminium Window Association	Suites 323/324 Golden House 28–31 Great Pulteney Street London W1R 3DD
The Architectural Aluminium Association	11 Cleeve Cloud Lane Prestbury Cheltenham Glos. GL52 5SE

Index